How to do your Student Project in Chemistry

How to do your Student Project in Chemistry

F. H. Jardine

Department of Environmental Sciences
University of East London
UK

CHAPMAN & HALL
University and Professional Division

London · Glasgow · Weinheim · New York · Tokyo · Melbourne · Madras

Published by Chapman & Hall, 2-6 Boundary Row, London SE1 8HN, UK

Chapman & Hall, 2-6 Boundary Row, London SE1 8HN, UK

Blackie Academic & Professional, Wester Cleddens Road, Bishopbriggs, Glasgow G64 2NZ, UK

Chapman & Hall GmbH, Pappelallee 3, 69469 Weinheim, Germany

Chapman & Hall USA., 115 Fifth Avenue, New York, NY 10003, USA

Chapman & Hall Japan, ITP-Japan, Kyowa Building, 3F, 2-2-1 Hirakawacho, Chiyoda-ku, Tokyo 102, Japan

Chapman & Hall Australia, 102 Dodds Street, South Melbourne, Victoria 3205, Australia

Chapman & Hall India, R. Seshadri, 32 Second Main Road, CIT East, Madras 600 035, India

First edition 1994
Reprinted 1995

© 1994 F.H. Jardine

Typeset in 10/12pt Times by Cambrian Typesetters, Frimley, Surrey
Printed in Great Britain by Page Bros, Norwich

ISBN 0 412 58360 7

A Catalogue record for this book is available from the British Library

Library of Congress Catalog Card Number: 94–071192

∞ Printed on permanent acid-free text paper, manufactured in accordance with ANSI/NISO Z39.48-1992 and ANSI/NISO Z39.48-1984 (Permanence of Paper).

Contents

Preface

When I was first approached to write this book I remembered the students I had seen, vainly trying to come to grips with the simple mechanics of producing a project report. Usually the chemistry itself had presented few problems, although sometimes it was clear that they had made an unfortunate choice of project. In other cases their endeavours could be likened to pouring $100\,cm^3$ beakers of water into the desert sands since they neither had any idea of the direction their project should take nor were they capable of steering it in that direction. Equally pitiful, as the deadline approached, were the sufferings of those who believed they could do their project in their spare time. Yet others came to grief upon obstacles they should have seen from a long way back.

There have also been students whose lectures on their work were distorted by speaker's nerves and whose distress has made me offer some advice on presentation and self-confidence. In fact few of their lectures were particularly bad, the really bad lectures more usually stemmed from overconfidence.

It is in the hope of preventing these and similar disasters that this book has been written. Students will look in vain in its pages for advice on the operation of the latest X-ray diffractometer or multinuclear NMR machine, even the operation of the pipette and burette has been virtually ignored. What I hope the readers will find is plenty of useful advice on the largely non-chemical aspects of carrying out and reporting on project work. It is possible to find some of this advice elsewhere. What I believe is the particular merit of this book is the variety of aspects that are covered in sufficient detail to enable students to perform to the best of their ability in project work. Its sole purpose is to help students complete their projects successfully with the minimum of trauma.

I have tried to deal with all aspects of project work from selection to the utilization of the knowledge gained in completing the project. Along the way readers will find such diverse topics as chemical safety in a research environment and advice on how to employ a typist.

When I was a callow youth I was arrogant enough to think I had made all possible mistakes in project work myself. However, after over a quarter of a century of teaching chemistry in universities in the United Kingdom and abroad I continue to be amazed at the sheer originality and creativity of the mistakes that are still perpetrated in project work. As others have remarked, the trouble with fools is that they are so ingenious. It is because of their ingenuity and the infinite variety inherent in projects and project work that it is impossible to make a guide like this truly comprehensive.

Any pretensions the guide has to being comprehensive have been largely due to the help I have received from others. Many inelegancies and ambiguities in the text have been removed by the watchful pencil of Richard Hubbard. Kathy Reynolds imparted much valuable advice on manuscript preparation and the trials of the typist in the course of a discussion that I hope was as cathartic for her as it was helpful to me.

I am greatly indebted to the staff of the Royal Society of Chemistry library for the help they have given me. Chapter 3 would have been much slimmer and poorer without their advice. Stuart Allen showed me just how many obscure control codes are required where a writer needlessly drives a word processor to the limit of its ability. Similarly Mike Cochrane ensured that the spectra shown as examples were indeed the best and worst of which the instrument was capable.

I would also like to thank not only all those of my former project students who produced excellent reports, but also those students whose early drafts alerted me to innumerable species of mistakes that can be made by the unwary and inexperienced. Unfortunately the student who was brave enough to give me an insight into her terror of public speaking cannot be thanked by name.

I had the foresight to seek the students' viewpoint of the text from Paul Mabbott and Wayne Jarvis who both went on to gain first class honours degrees. Neither has my own son Christian been averse to putting forward the current chemistry student's viewpoint when reading the draft manuscript.

However, as always, any remaining errors and omissions are mine alone. It will doubtless give observant readers particular pleasure to bring these to the attention of one who is purporting, at least in part, to be writing a guide for scientific authors. May I assure any amateur proof readers or sub editors lurking out there that it will give me equal pleasure to be informed of these so that they can be corrected in any future printings or editions.

Why projects? | 1

Nearly every undergraduate studying chemistry is now given a project to complete during the final year of the course. Chemistry projects are also playing an increasingly important part in taught Master's degree courses. Why are projects so popular, and what purposes do they serve in chemical education? Why have projects virtually replaced structured final year laboratory courses in several countries? What do you gain from completing a project instead of continuing with a standard laboratory instruction programme?

The cynical may say that a final year project represents the ultimate expression of the heuristic principle in education. The initial application of this principle had you making mud pies in your nursery school or kindergarten. However, there is much to be said for the process of 'learning by experiment' in your final year. At one end of the scale it represents a more or less orderly transition from the world of learning to the world of work. At the other end it can be your initial contact with the excitement and frustrations of scientific research.

The demise of the traditional laboratory course in final year chemistry degree programmes is a natural consequence of the rising importance of instrumental methods in chemistry. It is no longer possible either on economic or educational grounds to teach every undergraduate every modern laboratory technique when each technique now has its own specialized instruments. This is even more true at a time of rapid advances in instrument sophistication, performance and cost.

As your project progresses you will encounter a number of challenges. You will be required to meet these by applying new skills and techniques. Your responses to these challenges will be spurred by the knowledge that you are solving real problems for a real purpose. Obviously most challenges will vary from project to project but some are common to all projects.

You might reasonably expect to improve your skills in most of the following 14 areas in a typical project.

- Literature searching
- Devising experiments
- Planning the conduct of experiments
- Advanced experimental techniques
- Safe working practices
- Using advanced instruments
- Testing hypotheses
- Time management
- Interpersonal skills
- Meeting deadlines
- Writing scientific papers
- Editorial tasks
- Preparation of artwork
- Lecturing techniques.

It is quite possible that only two items (advanced experimental techniques and instrument operation) from the above list would form part of a traditional laboratory course. Thus, you can see that projects can be a very effective way of widening your experience and introducing you to new skills and techniques. Additionally, many of these skills will be of use to you no matter what career you may choose. The versatility you gain during your project will improve your employment prospects. Even your project report can be used to give a prospective employer a better indication of your abilities, particularly as your report may be available before your degree results.

Perhaps the most valuable aspect of project work is the foretaste it gives you of a career in either research or industry. As a result of your project experience you will be able to make a more considered decision about whether your career should be at the laboratory bench or the merchant banker's desk.

Choice of project 2

Very few universities allocate projects arbitrarily or by lot on the grounds that these are the fairest methods. In some cases your sponsoring company may have the option of choosing your project to fit their requirements. However, dictated assignments seldom allow students to achieve their maximum potential or to gain the maximum benefit from their project work. Most chemistry departments abide by the maxim of the early 19th century British navy and believe that one volunteer is worth ten pressed men. In these departments the students are allowed to select their projects from an approved list of titles, or, very rarely, to suggest their own project title. Some of these projects may be based outside the department either in industry or abroad.

However, the opportunity to choose your own project requires you to make your choice wisely. Your choice should not be made hastily, neither should it be based on trivial criteria. You should not, for example, sign up for the first project on the list that appeals, neither should you make a selection merely to avoid a particular supervisor.

The project should be selected only after careful evaluation of the varied aspects of project work that can contribute to a project's success. Important considerations are the feasibility of the project itself and how it relates to the time and facilities available for its successful completion. Consideration must also be given to the supervisor's expertise and whether the project plays to your own strengths and interests. These factors are discussed in more detail below.

2.1 PROJECT TITLE

A list of project titles and their supervisors posted on the departmental notice board is usually your first indication of the projects that will be on offer. Frequently these titles are accompanied by a brief description of the individual projects. You should read the list of titles and other information carefully before noting down any titles that initially appeal to you. Make sure that you are aware of the closing date for submitting your choice of projects.

You should make preliminary investigations of at least three titles since you may not be allocated your first or second choice. Placing more than six titles on your preliminary list may prevent you from investigating any of them sufficiently thoroughly and thus be counter-productive.

While there is nothing wrong in confining your selection of titles to one of the principal sub-divisions of chemistry (i.e. analytical, inorganic, organic or physical), a further self-imposed restriction to a more limited field within one of these subdivisions (e.g. steroid chemistry or zirconium(IV) complexes) must be avoided. A narrow range of choice does not augur well for your overall performance in your forthcoming final examinations. Never neglect the opportunity project work gives you to broaden your horizons.

Ideally your preliminary list should include titles from two of the principal subdivisions. Interesting titles can often be found on the borders of the principal subject areas. One example of such a title might be *The Synthesis and Coordination Properties of Organic Macrocyclic Ligands*.

It is also a good idea for you to investigate a 'wild card' title. If nothing else, investigation of a title away from your main interests will provide you with a better perspective on those topics with which you are more familiar.

Some project proposals are very detailed and closely argued. These projects, where it appears that nothing has been left to chance, may not appeal to the more adventurous. Nevertheless, provided there are no hidden shortcomings, they should give rise to very sound if unspectacular projects. Avoid projects with precise titles as they leave you with no flexibility. They prevent you from following interesting leads that open up as you develop the project and thereby reduce both your contribution to the project and any marks you may be awarded for original thought or initiative. More to the point, a precise title prevents you from changing direction should things go wrong.

A good example of a restrictive title might be *The Determination of the X-ray Crystal Structure of Ferrocene*. If crystallographic grade crystals of ferrocene cannot be obtained, the whole project falls at the first hurdle. In fact, perfect crystals of ferrocene have been sought without success for over 30 years. If your supervisor can convince you that the elusive perfect crystal of ferrocene is available for your exclusive investigation, then you have a world-beating project. However, even if such good fortune were to be yours, it would still not be the best project for you if you suffer from computer phobia nor if the scientific love of your life is natural product chemistry.

The titles of feasible, flexible projects which you stand a good chance of guiding to a successful conclusion are usually open ended. Some could even be described as vague. However, you should equally avoid excessively vague titles as these may reflect insufficient thought or a 'Mr Micawber style' attitude on the part of the proposer.

Examples of promising titles might be *Sydnone Synthesis* or *The Rates of some Cannizzaro Reactions*. Even the dreaded ferrocene project might be of more promise if entitled *The Crystal Structure of an η^5-Cyclopentadienyl Complex*. If no suitable crystals of ferrocene could be obtained, you would then be able to investigate any other η^5-cyclopentadienyl complex whose crystal structure had not been determined, without straying outside the confines of the title.

Many students select projects because of their inherent appeal. It is certainly easier to maintain your commitment and enthusiasm because you have a long standing interest in the project's area. For example, those with an interest in fashion may gravitate towards a project in dyestuffs. Silver(I) chemistry may be the subject for photographers, while motoring enthusiasts may select a project in fuel additives.

A more detailed investigation of your selected titles will normally begin with an informal discussion of the outline plans of work with the projects' proposers. Each title is normally supervised by its proposer. The supervisors will tell you what they wish you to achieve while carrying out their projects. In turn you should question them on the previous work that has been carried out in this area. You should try to obtain a list of references to this prior literature. By reading these references you will begin to gain some idea of the difficulties and pitfalls that may await you. This closer study should also reveal if the project really was that given in the title and resumé.

Important aspects of the projects that may start to become apparent at this point are the quantity and quality of the results you are likely to obtain. Will you have enough original material to write a report without having to pad it out? Will you be able to speak at some length on the results and their significance? In particular this more detailed scrutiny should leave intact your belief that the project is feasible.

Assuming that the project is still worthy of your consideration you should proceed to the next stages of the evaluation. These include the facilities available for the projects and the roles played by the supervisors in them.

2.2 SUPERVISION

Don't ever work for that slave-driving illegitimate X!' may be the conventional undergraduate wisdom in the department. However, in research just as in sport, nice guys have a habit of coming last. Successful research supervision often requires very different qualities from student counselling, and Prof. X may be just the person to provide the drive and initiative required to revive a flagging project.

Many lecturers exhibit very different characters when supervising research projects than when lecturing to large classes or even tutoring small groups. Your fellow undergraduates may be guilty of reaching unwarranted conclusions if they base their opinion of academic staff merely upon the demeanour of the latter in classes. Nevertheless, the performance in the lecture theatre can give some indication to the possible research performance of the supervisor. Are the lectures well organized? Has the information imparted been culled from a variety of sources? Is it up to date? Positive answers to these questions make it more likely that the lecturer will make a good supervisor. On the other hand, if the lectures are a jumble of disconnected facts from which you have great difficulty in identifying, let alone extracting, the main points then it is more than possible that the research will similarly lack direction.

However, a good lecture performance does not necessarily guarantee a flawless research performance, since original thought is not an essential requirement of a good lecture.

You may have already gained some indication of the effectiveness of your potential supervisors' writing style when you read the project topic's literature (Section 2.1 above). If none of the supervisor's work appeared there, then you should seek out some recent publications of the supervisor and investigate their quality. It is very important that your supervisor can write well. A large proportion of your project's marks will be derived from the written account you produce. You will be heavily reliant upon your supervisor's advice and literary skills in producing your final written report.

Your supervisor should be able to help you develop an economic and effective scientific writing style. You should investigate your supervisor's published work to see if it is also effective. Can you understand the main points your potential supervisor is trying to convey in the publications? Despite your limited experience, this should be possible if the paper is well written. Supervisors who are unable to write well themselves are unlikely to be able to impart the necessary skills to their students.

The people most critical of the academic staff's performance as supervisors are their own postgraduate students. It is worthwhile talking to these postgraduates to seek their views on the skills and attitudes of various possible supervisors. You might ask them how tolerant these supervisors are of mistakes or accidents. It is also worthwhile raising the question of their views on timekeeping and overtime working.

The postgraduate students will also be able to tell you about the progress of the different supervisors' research students. Supervisors whose post-graduate students complete their research degrees before the grant money runs out are more likely to see your project is finished on time than those whose postgraduates take twice the normal time to complete their PhD programmes.

How much experience does your supervisor have? Brand new assistant lecturers may be short on experience but long on new ideas. Do not necessarily believe that you should not go into the water before you can swim. The newer members of staff should still be able to remember first hand the problems students encounter during their projects. They may still be sufficiently idealistic to ensure that you, as a representative of the next generation, will not also suffer from these problems.

On the other hand you may think that it is the pioneers who are shot by the Indians, and may avoid new members of staff on principle. By doing so you may lose on enthusiasm whilst gaining on experience. New lecturers who are too arrogant or too shy to seek advice from their more experienced colleagues should certainly be avoided. Some new lecturers may feel they have to prove themselves in the eyes of their new colleagues. They will probably supervise your project very closely and lack the relaxed and more philosophical attitudes of more established members of the department. This approach may suit the more hesitant student but discourage the more originally minded.

Nevertheless, choosing a supervisor has much in common with choosing a lover. In both cases you will be spending a fair proportion of each day in each other's company. A supervisor who works from 4 pm to 2 am is as incompatible with the ordinary 9 to 5 person as one who starts work at 6 am. Sometimes incompatibilities are more productive than compatibilities. An optimistic supervisor can counteract a pessimistic student, but the reverse combination is less likely to be successful.

Irritating personality traits on both sides can sour the relationship between you and your supervisor. Unfortunately these often only become apparent when it is too late. Indeed, your project may be your first opportunity to develop that most important interpersonal skill – getting on with the boss.

The ideal supervisor is a leader rather than a driver. Those who think laterally are more amenable and productive than supervisors with tunnel vision. Enthusiastic supervisors inflict more wear and tear on their teams than the indolent, but are more likely to bring their projects to successful conclusions. Supervisors who allow their students to exercise their initiative rather than making them feel just another pair of hands usually gain more respect and support from their students. Equally, students should welcome the opportunity to contribute more to the project than just their labour, particularly where they are rewarded for so doing.

Ultimately, however, selecting a supervisor is a personal matter. It involves such intangible factors as trust and confidence besides the less subjective assessments of the supervisor's competence and knowledge. In the end it comes down to a belief that the supervisor is in charge of a project that you can jointly steer to a successful conclusion. If you do not take offence at some of your rough corners being removed along the way it may even be fun too.

2.3 THE TIME AVAILABLE

The total time available for the project is of great importance in determining your choice, but the quality of this time is even more important. The quality of time is measured by how it relates to the tasks involved in the project. To take the most trivial example, short laboratory periods are incompatible with long reaction times.

The supervisor should have made some estimate of the total time required to achieve the aims set for you. However, the corollary to 'nothing is as simple as it looks' is 'everything takes longer than you think', and the supervisor may be expecting the performance of an experienced post-doctoral student from a relatively inexperienced undergraduate. You might ask each supervisor what allowances have been made for your inexperience in arriving at the aims of the project. Despite any assurances you may receive, it is up to you to make your own estimate of the time each possible project would take to complete.

Possibly the most trivial factor in these estimates is the actual time the experimental work will take. As was outlined in Chapter 1, projects involve much more than just working at the bench. Nevertheless, you should make an estimate of the time required at the bench as a first check of the project's feasibility. Obviously the total synthesis of a rare steroid from three-carbon building blocks cannot be achieved in two weeks. Neither can you expect to master the operation of a complicated gas chromatograph working in tandem with a mass spectrometer in this time. The slow reactions of complexes of the third transition series elements may require much more time to complete than cognate reactions of second transition series species.

As you will find to your cost, the time taken to carry out the experimental work is usually less than the time taken in preparing to carry out this work. This is apparent from some of the factors shown in Table 2.1.

Delays eat into your project time from the very beginning. Your literature search is first slowed by the journal you require not being on open access in the library. When you eventually receive it from the counter you find that the key reference it contains is not held by your library and a copy must be ordered from another library. Unfortunately when this copy arrives it is in Russian and you have to spend further time and effort obtaining a translation. Thus, a task that might have taken you five

Table 2.1 Factors determining the effective project time available

Division	Constraints
Literature search	Opening hours of library
	Access to journals
	Availability of obscure journals
	(i.e. the speed of the interlibrary loan service)
	Translation time and facilities
Experimental work	Availability of chemicals
	Obtaining chemicals from manufacturers
	Total laboratory time allotted
	Length of laboratory periods
	Instrument time allotted
	Analysis turn-round time
Writing up	Rate of accumulating results
	Supervisor's approval of draft
	Typing time
	(a) typist availability
	(b) word processor availability
	Preparation of artwork (if any)

minutes under ideal circumstances ends up taking five weeks. The problems encountered in the literature search are likely to be similar in all projects. However, if nearly all the references you receive in the supervisor's initial list are to obscure Chinese journals and you cannot read Chinese, you may be able to use your time more efficiently in a project where the relevant literature is mainly confined to recent issues of the *Journal of the American Chemical Society* or the publications of the Royal Society of Chemistry.

To carry out the experimental work you will need to use chemicals. Are these chemicals available in the departmental store? If you have to ask your supervisor to order them from the supplier, how long will it be before you receive them? What is the departmental policy on ordering chemicals? Are they ordered as required or at weekly or even monthly intervals?

You should consider the availability of chemicals and the departmental policy as a minor factor in your choice of project. If your department holds a limited range of chemicals and has a restrictive ordering policy then this must favour a project using simple, available chemicals. In may be possible to save time by either reserving chemicals in advance or by placing advance orders through your supervisor. If your project is selected before the long vacation, then ordering your chemicals for the next semester before you depart on vacation should certainly save time.

It is particularly important to avoid delays at the start of the project, since time wasted then cannot readily be retrieved later. Similar delays at a later stage can be used to carry out some other project-related activity. This might be preparing a first draft of the report or running a spectrum you have overlooked.

The importance of the maximum length of laboratory periods has been noted above. There should be sufficient time to carry out long reactions or to complete kinetic runs. Reactions may be allowed to continue overnight at the expense of lower yields, but kinetic experiments must have sufficient time for setting up and completion in a single laboratory session. Your preliminary investigation of the project should aim to discover any slow reactions in either kinetic or preparative projects.

There is no point in having long laboratory sessions if there is a limitation on the time you may spend using an instrument. Make sure that you have sufficient access to any instruments which you may need. You may be competing for instrument time with postgraduate students, so you should establish with your supervisor the priorities allocated by the department.

Beware of projects located outside your own department. It is very likely that you will have a very low priority on another department's equipment. If you are using facilities outside your university be sure to allow for travelling time when considering the time available to you.

It may be that you will find that certain instruments are not available for direct undergraduate use, but are available as a departmental service. If these services are required in your project ask your supervisor what restrictions are placed upon them. For example, how many CHN analyses will you be allowed? What is the turn round time for samples run as a service on the 300 MHz NMR machine? Can you make prior arrangements for unstable samples to be analysed by appointment?

It is never too early to start writing up your project, and writing up small pieces as you go along is an excellent way of making efficient use of your time. As will be explained in Chapter 6 it is often easier to write the experimental section first. Practise by writing up your laboratory notebook in the style you will use in your report. You can then easily make a first draft of the initial work for your supervisor's approval. The supervisor's corrections will place you on the right road and save much wasted time later. It will also give you an indication of how rapidly your supervisor will return your work later in the project when the heat is on.

Be sure that you have allowed sufficient time for the preparation of the final copy of your report. Remember that you will either have to prepare a legible draft for your typist, or type out the report yourself. If you select the first option this will mean engaging a typist in good time and finding out how long it will take for your report to be typed professionally. With the second option you must calculate how long it will take you to type the report. The rate-determining step may be the availability of a word processor. In both cases it is vital that you allow sufficient time for correcting the final product. You must also allow time for the preparation of artwork for your report. This artwork might include display equations besides the more obvious figures and diagrams.

2.4 THE FACILITIES AVAILABLE

It should be self-evident that the facilities required to complete your project must be available to you. While it is relatively simple for you to check on the physical presence of instruments, you should investigate further those instruments you may wish to use. Instruments which have lain dormant since the previous project student used them a year ago are unlikely to be in top-class working order. They may require servicing before you can rely on the results they produce.

You should inquire about the departmental policy on instrument servicing. Ask how long a vital instrument might be out of service awaiting repair. Can the department repair most of its own instruments, or have service contracts been made with the manufacturers? Can the department always afford to have its instruments serviced immediately? Perhaps the postgraduate students are the best people for you to question in this vein. They will also hold suitably jaundiced views on the reliability of certain instruments.

From this information you will be able to decide if a project which relies heavily upon a few instruments is a desirable choice. Unreliable or unusable instruments may enforce a major directional change in your carefully devised project. In emergencies it may be possible to seek help outside the department, but this depends greatly upon the facilities available in the immediate area. As noted above, you may find that you are accorded low priority on outside instruments. You may find that instead of investigating the sample yourself it will be run as a service elsewhere.

By all means make use of the instrumental services that may be available unless there is an advantage to be gained in investigating your own samples. For example, carrying out your own CHN analyses is not usually considered to be a major contribution to a project in preparative organic chemistry. On the other hand, if you are investigating the infrared spectra of weakly coordinated polyatomic anions it is essential that you carry out the infrared spectrometry yourself.

Generally speaking you should not spurn the opportunities a project provides to learn to operate an instrument. This specific knowledge may give you a small advantage when seeking employment as well as generally increasing your familiarity with instruments in an increasingly instrumented world. However, you should not forget that obtaining the best perform-ance from an instrument takes time, skill and patience.

2.5 MAKING YOUR FINAL CHOICE

You should now have examined in some detail the important components of several projects. Before you can make a final choice all these components must be evaluated for each project. This can be done by giving a score, say out of ten, to each component and then multiplying it by a weighting factor from 1 to 10 according to its importance in the overall scheme of things (Table 2.2). Obviously the weighting for each aspect will vary from project to project. The importance of the supervisor in a difficult X-ray crystallography project is greater than that of the supervisor in a routine project in organic synthesis. Conversely, chemical availability may be weighted more highly in the latter project. The project gaining the highest score should be the most suitable.

Table 2.2 Assignment of weighting factors to decision components

Component	Score	Weighting factor	Product
1. Overall appeal			
2. Scope for originality			
3. Flexibility of topic			
4. Sufficient results for report			
5. Accessibility of literature			
6. Compatible supervisor			
7. Reasonable expectations			
8. Expertise available			
9. Short induction period			
10. Instrument availability			
11. Chemical availability			
12. Realistic timescale			
13. Experimental complexity			
14. No external facilities required			
	Total		

Unfortunately you will probably find that two projects have nearly equal scores. In these cases you should tinker with the weightings you assigned to see if this gives a clear victory to one or the other. If they are still equal or nearly equal at this point then follow your instincts when making your choice.

Once your choice has been made or your project assigned there can be no going back. You must concentrate all your energies on what is now **your** project and not look back wistfully on the project that might have been.

Using the library 3

Your project is often used to teach you how to use the full range of library services available in the place of an artificial exercise. Knowledge is strength in all types of project and this knowledge can most easily be acquired by using the resources of your library. It is probable that during your undergraduate career you have mainly confined your reading to textbooks or what is known as the tertiary literature. Your project offers you the opportunity to become acquainted with abstracting journals, the secondary literature of reviews and monographs, besides reading the primary literature of original research articles in the chemical journals of the world.

The chemical literature is vast. Currently, over half a million chemical articles are published in the world each year. It is therefore impossible for anybody to keep abreast of all the literature by reading over 1000 articles a day. Indeed, it is becoming increasingly difficult to keep abreast of all the articles published in a selected field. Authors of wide-ranging review articles are beginning to find that new material appears faster than they can include it in their text. This results in reviews becoming outdated before they appear in print.

The only way to cope with this deluge of information is to be selective. One unfortunate consequence of the literature explosion is that the proportion of really significant articles declines each year. Thus another valuable part of your training is learning to recognize and reject the dross, since an exhaustive search will undoubtedly leave you too exhausted to contemplate laboratory work.

3.1 SEARCHING THE LITERATURE

At the beginning of your project you will need advice on where to start fishing in this vast ocean of chemical literature. Your supervisor, as noted in Chapter 2 above, should have referred you to some of the more important papers in your project area when you were making your preliminary investigation of your project. Using these papers as a starting point, it is quite possible to make a modest, retrospective literature search by following up the references they contain. However, this is not a particularly efficient method of conducting a literature search. You may gain some further information about your project topic by finding and reading specialized books, where these exist. It is pointless to go to the library and pick up journals at random and read them in the hope of finding something relevant to your project. The chemical literature is much too large for this approach to stand any chance of success.

The key to unlocking the chemical literature is *Chemical Abstracts*. This huge work, dating back to 1907, aims to provide an abstract of every chemical article published in the major scientific and technical journals of the world. Relevant books, symposia proceedings and dissertations are also abstracted. Abstracts of the chemical patents lodged in the world's industrialized countries form a large proportion of *Chemical Abstracts*.

Parts are published weekly and each currently contains about 10 000 abstracts. The abstracts in each part are grouped into broad subject areas and then further subdivided within these areas.

Each abstract (Figure 3.1) is assigned a unique reference number prefaced with its volume number. Besides summarizing the contents of the article, the abstract also gives the article's location to allow you to seek out the more detailed information it contains.

The strength of *Chemical Abstracts* is not only its comprehensive coverage but also its good indexing system. Each article abstracted is indexed by author(s), subject, chemical substance and chemical formula. Patents have their own index. The weekly parts each contain a keyword index. Main indices are published twice yearly to correspond with the

93: 94418n Mechanisms of 1,1–reductive elimination from palladium. Gillie, Arlene; Stille, J. K. (Dep. Chem., Colorado State Univ., Fort Collins, CO 80523 USA). *J. Am. Chem. Soc.* **1980**, 102(15), 4933–41 (Eng). Three cis Pd complexes [e.g. $(PPh_3)_2PdMe_2$] underwent reductive elimination in the presence of coordinating solvents (Me_2SO, DMF, THF). The trans complexes which could isomerize to cis did so in polar solvents and then underwent reductive elimination. The eliminations from the cis isomers were intramol., as detd. by the lack of crossover with the perdeuteriomethyl Pd analogs and displayed first-order kinetics.

Figure 3.1 An entry from *Chemical Abstracts*. (Reprinted by permission. Copyright 1980 by the American Chemical Society.)

biannual volumes of *Chemical Abstracts*. Originally these biannual indices were collected into decennial indices, but the quantity of material is now so great that quinquennial indices have become the rule. You will save time by searching the collective indices rather than the biannual indices. The more recent work will, of course, be found in the biannual indices as the collective indices take time to compile.

To aid your search of the indices *Chemical Abstracts* publish an *Index Guide*. This helps you find the most useful terms for your search. It mainly consists of the synonyms of chemical names. These can occur when trade names are in common use. For example Freon 123 is indexed under Ethane, 2,2-dichloro-1,1,1-trifluoro. Synonyms also arise when a system of nomenclature has been revised. However, the newer system does not always replace the old. The newer names of bacteria are used in the index (e.g. *Clostridium welchii* has been replaced by *Clostridium perfringens*), but you will look in vain for ethanoic acid in the indices. Fortunately the *Index Guide* comes to your rescue by stating 'see Acetic acid *[64-19-7]*'.

Chemical Abstracts was one of the first abstracting systems to make use of modern information technology to allow computer searching of its database. All abstracts published since 1967 are in this database. Your university library may have access to this database. Failing this the Royal Society of Chemistry, the Patent Office and some commercial firms can also search the database. A word of warning, however: charges for using the system have increased greatly in recent years and you must obtain authority before using the system. One thousand references turned up at 10p per reference gives a bill of £100 before connection charges! The Royal Society of Chemistry currently has a minimum charge of £30 (+ VAT) for the service.

Possibly your university library has negotiated preferential academic rates either by using the system at non-peak times or by having the output printed off-line. By exercising a little forethought and deviousness you may be able to have your literature search performed free as part of the library's demonstration programme. The best way of keeping charges to a minimum is, like manual searching, to define your search as precisely as possible. In both cases the wider your search the more references you will need to read and you will have more decisions to make before including or rejecting them. However, it is good practice to make your search slightly wider than you require to ensure that you have not missed anything of importance.

Let us suppose that you are searching the database for papers dealing with the methyl complexes of the platinum group metals. To instruct the computer to record all papers where the word 'methyl' appears would probably cause it to regurgitate about half the contents of *Chemical Abstracts*. This output will, of course, include innumerable, irrelevant references to methylamine, tetramethyltitanium (IV) and all the other methyl compounds in addition to those actually sought. Limiting the search term to 'methyl complexes' would avoid listing the methylamine references but would still include the unwanted references to the titanium complexes.

You may possibly miss many important references to complexes not containing platinum if the computer only lists those abstracts where **both** 'methyl complexes' and 'platinum' occur despite you specifying 'platinum group metals'. A better strategy would be to ask for references where the keyword 'methyl complex' occurred in conjunction with 'ruthenium', 'osmium', 'rhodium', 'iridium', 'palladium' or 'platinum'.

The search program is very powerful and if, for example, one paper dealt with the methyl complexes of both osmium and palladium it would only appear once in the printout. The search may be narrowed further if you were only concerned with the decomposition or catalytic properties of the complexes. Instructing the computer to report only those abstracts where one or more of the six platinum group metals, 'methyl complex' and 'decomposition' all appeared will reduce the quantity of abstracts.

If only the catalytic properties are sought, a further feature of the search program should be used. The keyword 'catalytic' may not include all the relevant material since some papers may be indexed under 'catalysis by' or 'catalyst'. You should make the additional limiting word 'cataly?' to embrace all the above possibilities.

The abstracts dealing with specific compounds can be sought by using their unique *Chemical Abstracts* registry number (a few substances whose names are used ambiguously may have more than one registry number). It is also possible to search for the work of a particular author. The search can also be restricted to any particular time span after 1967.

It has been the author's experience that the system is very good but not infallible. About 1% of abstracts have been mis-indexed. This means your output will contain about 1% irrelevant abstracts and about 1% of relevant abstracts will be omitted. Many errors arose in the early years when bored undergraduate piece-workers entered the backlog. One of their favourite tricks was to omit one letter from the keyword 'lanthanide shift reagent'. Searching the database also reflects the main shortcoming of the *Chemical Abstracts* system which is very good at unearthing facts but very poor at listing ideas.

The output can be received in two forms. The first, and more expensive, prints out the title, authors, and location of the original paper. If you have only a few references this method may as well be used. It can also be used as an initial check on the accuracy of a larger search. Economy dictates that large searches only print out the abstract number and the volume of *Chemical Abstracts* in which it appears. You can then look up the abstracts reported and make a preliminary selection from them. If the abstract appears sufficiently relevant for the original to be consulted then you should record the names of the authors and the paper's location on a reference card (Section 3.2).

Manual searches follow a similar procedure except that the abstract number is obtained from an index volume. Note that pre-1967 indices refer to columns in *Chemical Abstracts* and give a letter to locate the abstract more precisely. Always note both the volume and abstract number on your reference cards to guard against transcription errors.

The *12th Collective Index* to *Chemical Abstracts* is now available on CD/ROM. This can be searched in a similar manner to the computer database. Several strategies can be adopted. If a particular compound is sought, the most rapid method is to use its *Chemical Abstracts* registry number. Alternatively it can be sought by entering its empirical formula.

Suppose we wished to find out if further work had been carried out on the complex $[PtMe_2(PPh_3)_2]$. We could enter its registry number, *[17567-35-0]*, and obtain the three abstracts making reference to it, namely 109:211201, 114:247494 and 115:293315. Alternatively, we could enter its empirical formula in the style $C_{38}H_{36}P_2Pt$, whereupon the system would respond with its registry number and the same three abstract numbers. Any other compound can be found by entering its empirical formula in the order carbon first, hydrogen second, and followed by the other elements in alphabetical order. In this system NaCl would be ClNa, and K_2WO_4 would be K_2O_4W.

You will recall that a favourite 'A-level' question was 'How many different compounds have the empirical formula . . . ?. Entering the empirical formula $C_{36}H_{38}O$ turns up abstracts for 12 different compounds. Hence the empirical formula method is much less useful for organic compounds of low or medium relative molar mass.

These methods of searching are also inapplicable when seeking information about very important or very common chemicals. The system cannot cope with the deluge of abstracts that are given for these species, and will ask you to refer to the chemical substance index.

In turn the chemical substance index is very inefficient in tracking down a few abstracts. If you searched for $[PtMe_2(PPh_3)_2]$ by this method you would have to go through the stages:

1. enter primary keyword – PLATINUM
2. enter secondary keyword – SUBSTITUTED
3. key through several thousand entries to reach 'dimethylbis(triphenyl-phosphine)'.

During the lengthy stage 3 you would have plenty of time to reflect how far artificial intelligence is behind human intelligence. A human of only average intelligence conducting a manual search would turn over ten pages at a time to reach 'dimethyl' more quickly. Still, it could have been worse: you might have been searching for a platinum xanthate!

One advantage of the CD/ROM system is that it is failsafe. If you enter a non-existent registry number or an improbable empirical formula like

$C_{100}H$ then the system reports 'No match found'. If you enter the stupidly impossible name 'pentaiodoacetic acid' you are referred to the nearest match which is 'pentane' in this case.

Organic chemists in particular have alternatives to using *Chemical Abstracts*. If you merely wish to know the physical properties of an organic compound, then it is probably quicker for you to look in the *Dictionary of Organic Compounds* and its supplements than to consult *Chemical Abstracts*. There are also smaller dictionaries listing the properties of organometallic or organophosphorus compounds. The *Handbook of Data on Organic Compounds* is perhaps less convenient to use than the *Dictionary of Organic Compounds* but it provides numerous references to the spectral properties of organic compounds. The *Merck Index* is also a useful source for details of the preparation and properties of pharmaceutical compounds.

The 'Rubber Bible', more correctly known as the *CRC Handbook of Chemistry and Physics*, is published annually and is a mine of information on nearly every aspect of chemistry. It is more a reference work itself than a source of references. It is ideal for a rapid survey of physical properties of groups of compounds. However, the table entries in this work do not always follow the standard conventions of chemical nomenclature which can make it difficult to use.

More detailed information upon a wider range of organic compounds appears in Beilstein's *Handbuch der organischen Chemie*. To use this you will have to both understand scientific German and crack the code by which compounds are assigned to sections. Guides to using Beilstein are available. Although scientific German sometimes seems a different language to literary or conversational German, using a scientific German dictionary intelligently can prove to be the key that unlocks the way to essential information on the preparation and properties of many organic compounds prepared around the turn of the century.

Unfortunately the corresponding reference work in inorganic chemistry, *Gmelins Handbuch der anorganischen Chemie*, is much less comprehensive. Although both handbooks make copious references to the early literature neither could be described as being universally up to date.

If you are more interested in the preparation of compounds, the series entitled *Organic Syntheses* or *Inorganic Syntheses* should not be overlooked as sources of tested preparations. The former, which is of earlier origin, also exists as collective volumes.

Review articles are another source of references to the primary literature. Many books and periodicals consist solely of review articles. Examples that spring to mind include *Chemical Reviews*, *Chemical Society Reviews* and the numerous titles in the *Progress in . . .* and *Advances in . . .* series.

During the last decade or so several series of reviews have been published under the general title *Comprehensive . . . Chemistry*. Topics covered include coordination complexes, organometallic compounds, heterocyclics, medicinal chemistry and polymer science. These series are useful sources of references, particularly to work carried out in the late 1970s and early 1980s.

There used to be an annual listing of review article titles published in analytical or organic chemistry, but this useful service appears to have been discontinued. Your supervisor may subscribe to a current-awareness-journal in the specialized field of your project. This is effectively a semi-personal abstracting service. Because of its more limited coverage it may abstract the primary literature more rapidly than the universal *Chemical Abstracts*.

Reviews and abstracts are also supplemented by *Annual Reports of the Royal Society of Chemistry* which refer to and comment on the highlights of the year's chemical literature. To some extent the twin volumes dealing with organic or inorganic and physical chemistry have become victims of the literature explosion. This has been recognized by the Royal Society of Chemistry, which also publishes more specialized works with the same aims. Examples of titles in these series include *Amino Acids and Peptides*, *Catalysis*, *Organometallic Chemistry*, *Carbohydrate Chemistry* and *Organophosphorus Chemistry*. There is also an annual review volume published as part of the *Journal of Organometallic Chemistry*.

Abstracts, handbooks, reviews and reports all enable you to work back through the chemical literature by following up the references each contains. You can even, if necessary, follow up the references in their references and go further back towards the source. However, there is another type of publication that enables you to search, albeit inefficiently, forwards in time from a key reference. This is the *Science Citation Index* which lists the citations of most scientific papers in annual volumes. The citations are arranged alphabetically by the first author of the original paper. Each author's publications are then listed chronologically. Recently a specialized offshoot, *Chemical Citation Index*, has appeared to fulfil this function in the chemical area alone.

The inefficiency of using the citation indices arises from the inevitable diversity of the work that follows. You will be very fortunate indeed if each citation of your key paper continues the original theme.

Whatever method you use to find references, make sure that once found they stay found. There are few more frustrating tasks in life than vainly searching the literature for a reference you have mislaid. Never make notes on odd scraps of paper like bus tickets or till receipts. Always use a notebook or a reference card. Ideally the latter should be stored in a box, but at the very least hold the batch together with two stout rubber bands. The cards' legibility is not improved by dropping them in the gutter, and 100 reference cards take an awfully long time to pick up on a breezy day.

3.2 REFERENCE COLLATION

After you have read the reference and decided to use it, take a reference card and make notes. A typical completed reference card is shown in Figure 3.2. The first line gives a general subject area so that it can be retrieved from the stack at the appropriate time when writing. This one has been assigned the working title 'Decomp. of methyl complexes'. A small annotation on the right shows that it was later assigned the temporary number of reference seven in your Discussion section.

The second line shows the names and initials of all the authors, as given at the beginning of the article, together with the name of the journal, year of publication, volume and page numbers. You should double-check these details against the original since they will be the only way you or anybody else using your work will ever be able to find the reference again. Even if you know that the 'A. Gillie' named on the paper was really 'A. L. Gillie' you must use the former name to agree with the original paper. The last entry on this line, [93-94418], is the *Chemical Abstracts* number, enabling you to refer back if anything has become garbled.

If the reference is very important you will certainly need to refer to it again in full when you come to write your report. This also provides a random cross-check on the accuracy of your references. Less important references may not need to be looked at more than once to abstract background information. The remainder of the card is used to make your

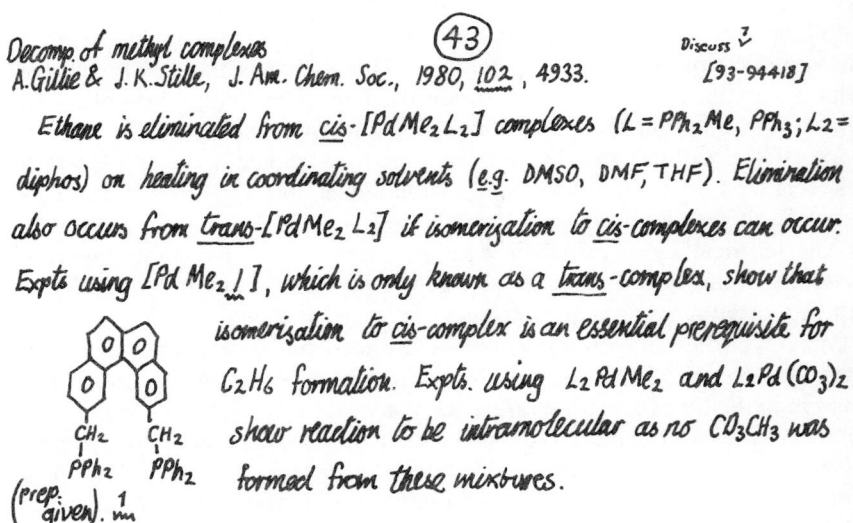

Figure 3.2 A typical completed reference card.

own abstract of the paper. Always make this abstract since it is good writing practice and it teaches you to be concise. Many students spend a fortune on photocopying and never develop their précis-writing skills. Psychologically, you also have better recall of something you have written than something you have merely read.

Writing your own précis has a further advantage. The précis can easily be incorporated in your own text since **you** wrote it. If you try to string together copies of abstracts made by a variety of authors, then your readers will be confronted with stark changes of style. This will tell the more intelligent ones that you are merely copying.

The abstract shown in Figure 3.2 does not follow the paper's abstract since in this case only ethane formation is of interest. At the initial reading you believe that a ligand which occupies *trans* positions in a square planar complex may also be of interest, so you note that its preparation is given in the paper. If you were certain that you will wish to prepare this ligand or a congener you could also note its preparation on the back of the card. In that case you would add a second working title to the card such as 'Ligand prep.'.

Your final application of the reference card is to use it to assign the reference's number in the master list. This reference finally became number 43 in the master list. By noting the master list number on the reference card you make it slightly less traumatic to include or delete references at a late stage. Once the master list has been finalized it can be typed out directly from your reference cards placed in numerical order. By assigning a master list number at the reference's first appearance in your final text you will avoid such solecisms as citing the same reference twice!

FURTHER READING

1. B. Livesy, *How to use Chemical Abstracts, Current Abstracts of Chemistry and Index Chemicus*, Gower, Aldershot, 1987.
2. J. Buckingham (ed), *Dictionary of Organic Compounds*, 5th edn, Chapman & Hall, London, 1982.
3. J. Buckingham (ed), *Dictionary of Organometallic Compounds*, Chapman & Hall, London, 1984.
4. S. Edmundson (ed), *Dictionary of Organophosphorus Compounds*, Chapman & Hall, London, 1988.
5. R.C. Weast and J.G. Grasselli (eds), *Handbook of Data on Organic Compounds*, 2nd edn, CRC Press, Boca Raton, FL, USA, 1989.
6. D.R. Lide (ed), *Handbook of Chemistry and Physics*, 74th edn, CRC Press, Boca Raton, FL, USA, 1993.
7. A.R. Katritzky and C.W. Rees (eds), *Comprehensive Heterocyclic Chemistry*, Pergamon, Oxford, 1984.
8. G. Wilkinson, F.G.A. Stone and E.W. Abel (eds), *Comprehensive Organometallic Chemistry*, Pergamon, Oxford, 1982.

9. G. Wilkinson, R.D. Gillard and J.A. McCleverty (eds), *Comprehensive Coordination Chemistry*, Pergamon, Oxford, 1987.

10. C. Hansch, P.G. Sammes and J.B. Taylor (eds), *Comprehensive Medicinal Chemistry*, Pergamon, Oxford, 1990.

11. G. Allen and J.C. Bevington (eds), *Comprehensive Polymer Science*, Pergamon, Oxford, 1989.

12. *Index of Reviews in Organic Chemistry*, Royal Society of Chemistry, London, 1971–85.

13. *Index of Reviews in Analytical Chemistry*, Royal Society of Chemistry, London, 1982–83.

14. S. Budavari (ed), *The Merck Index*, Merck and Co., Rahway, NJ, USA, 1989.

Safety　　　　　　　　　　　　4

Your project represents the transition from well tried and tested, inherently safe, elementary undergraduate laboratory experiments to real chemical experiments. In the former experiments, the apparatus will have been specified by the lecturer in charge and inspected before use by the technical staff. Now you must find the safest way to carry out the experiment. You must also select your apparatus and inspect it for flaws before commencing work.

You should always ask your supervisor about any possible hazards when devising experimental procedures. If you are working in an obviously hazardous area such as with nitro or diazonium compounds, polynuclear aromatics, azides or cyanides, it may also be desirable for you to seek the departmental safety advisor's advice and approval.

Valuable sources of information on chemical hazards are listed at the end of this chapter. You should treat all chemicals as potentially hazardous. However, for a chemical to cause harm it must come into contact with the body. Provided the precautions taken to prevent this contact are effective, then even the most hazardous chemicals can be used safely.

There is little point in you taking all possible precautions against the hazards of a particular chemical when it is in use, only for you to dispose of it in an unsafe manner at the end of the experiment. You must consider the hazards a chemical presents at every stage from acquisition to disposal. Before withdrawing the chemical from the store or ordering it from a supplier, its hazards should also be discussed with your supervisor.

The level of formal supervision of your detailed activities will be reduced, both on logistic grounds and as a concession to your greater experience. You will be approaching a situation where you are your own safety officer. Under these circumstances it is as well to lay down your own safety rules, in addition to any imposed by the department. Some of the more obvious points include:

Never work without another person within call
Always wear approved safety glasses in the laboratory

- Never wear contact lenses in the laboratory
- Know the locations of emergency exits and safety equipment
- Know how to use safety equipment (e.g. fire extinguishers) **before** an emergency arises
- Have antidotes and emergency equipment to hand **before** starting work
- Wear protective clothing and stout, sensible shoes in the laboratory
- Restrain long hair and flowing clothes
- Use the safety equipment provided
- Never smoke, eat, drink or apply cosmetics in the laboratory
- All containers must carry the name of their contents
- Injuries must be treated immediately, irrespective of seriousness
- All accidents must be reported to your supervisor
- Clear up spillages and broken glass immediately
- Treat all chemicals, particularly new or little known compounds, as hazardous
- Display appropriate warning notices for others' benefit
- Do not use the fume cupboards as a store
- Return all reagents to shelves or storerooms after use
- Never leave unlabelled material or residues for disposal by others when you have finished your project
- Know the meaning of the safety symbols you will encounter.

4.1 BECOMING YOUR OWN SAFETY OFFICER

The Health and Safety at Work Act came into force in 1974. You may be surprised to learn that you are outside the provisions of this Act. The Act only applies to employees, and students merely have the status of visitors, although some industrially sponsored students may come within the provisions of the Act. Nevertheless, this does not give students *carte blanche* to carry out any dangerous experiment they choose, since in law they still have a duty of common care imposed upon them. Your supervisor is, however, subject to the full rigours of the Act. For this reason alone most university chemistry departments operate as if the Health and Safety at Work Act did indeed apply to students.

This system makes the transition to a working environment much simpler for graduates. It is one of the advantages of project work that you will become acquainted with a wider range of experimental techniques and chemicals than you might in standard laboratory classes. What is more important is that you will be required to make your own investigations into the safety of your project's experiments.

Although hazards are more generally associated with the preparative aspects of chemistry, a carelessly handled gas-chromatography syringe can cause a nasty wound. It may also inject a toxic chemical into your tissues

Similarly, physical chemists should be aware of the dangers associated with mercury spillages, high voltage electricity, lasers and cryogenic substances.

3.1.1 Good housekeeping

Most accidents can be avoided by simple good housekeeping. You will be obliged to keep your own working area free from hazards. Additionally, you may no longer be working in an undergraduate laboratory where everything has to be cleared away at the end of each session or experiment. Instead you may find yourself assigned permanent bench space in a small research laboratory. This change of venue will call for a change in working practices on your part. Without the discipline of clearing away at the end of each session you may develop untidy habits to the irritation of, if not anger to, your immediate colleagues.

You will also have to carry out safely many of the routine operations normally performed in the teaching laboratories by the technical staff of the department. These tasks may include preparing your own solutions and filling reagent bottles. You should take care when dispensing and dissolving even such common reagents as sulphuric acid or sodium hydroxide. When filling reagent bottles remember to leave a 10% headspace for expansion. If the reagent is inflammable, avoid static electricity by earthing the container. Nylon clothing is a source of static electricity and, since it melts into skin burns, should never be worn in the laboratory.

Always wear stout leather gloves when trying to remove a stubborn stopper from a bottle or reaction flask. If the flask or bottle should break, the gloves will protect your hand. Gloves should also be worn when you are cutting glass tubing.

Suitable goggles must be worn if you are working with ultraviolet light sources or lasers. Goggles should also be worn when glass blowing. It is no coincidence that most glassblowers have to wear ordinary spectacles.

3.1.2 Overnight experiments

Ensuring continuity of supplies to experiments left overnight is very important. You should take particular care that the flow of water to condensers is not interrupted. You should wire all rubber hose connections in place and ensure that the hoses cannot become kinked or displaced from the sink. Never wire the rubber hoses in place with bare copper wire since this extracts the sulphur from the rubber and causes it to perish. A piece of thin lead sheet wrapped around the end of the exit hose makes it more difficult to displace from the sink. Test the system at higher than normal water pressure since the water supply pressure often increases overnight due to lack of demand. Water flooding from a defective condenser system can cause very expensive damage to the floors below. The inconvenience is immeasurable when paper records are turned to pulp by the water.

All experiments left on overnight must be clearly labelled, stating the reagents present, and a safe shut-down procedure in case of emergency. The label must also state where you can be contacted in an emergency. Most departments require the authorization of a member of the academic staff before any experiment can be run overnight.

4.1.3 Preventing explosions

There is the additional danger of an explosion if a reaction boils dry. Involatile hydroperoxides accumulate in the evaporating organic solvent and can detonate if subjected to further heating or shock. You should never distil to dryness in search of maximum yields.

Take particular care when working with inherently unstable compounds such as azides, diazomethane, polynitro-compounds, chlorate(V) and chlorate(VII) salts, manganate(VII) salts, concentrated hydrogen peroxide solutions and other powerful oxidants. Caution should also be exercised in reactions or decompositions that can release large quantities of gases, as these species or mixtures frequently form the basis for commercial explosives or detonators.

You should pay attention to the reactions in your charge at all times. Prior notice of an explosion is often given by changes in the appearance or temperature of a reaction mixture. The sudden appearance of fine cracks in a reaction vessel also signals imminent danger.

Beware of causing an explosion by igniting the vapours of flammable organic solvents. If you are extracting a mixture with a flammable solvent vent the separating funnel in a fume cupboard, or at least ensure that there are no sources of ignition present in the laboratory.

4.1.4 Vacuum lines

New hazards go hand in hand with new experimental techniques. A vacuum line will bring implosion hazards. It may also be a fire risk since many gaseous reagents that might be handled in a vacuum line are spontaneously inflammable (e.g. diborane). All large, evacuated vessels should be encased in a metal safety screen. Failing this a safety glass screen should be fixed in place between the vessel and the operator where this is reasonably practicable. Before turning vacuum line taps, always warm them with a hair drier to make the lubricant less viscous and reduce the strain imposed by turning them. Inspect vacuum lines regularly for any flaws. During these periodic inspections release and gently retighten clamps in turn to avoid any strain. Always allow room for contraction and expansion if the apparatus is to be subject to temperature changes.

1.1.5 Cryogenic substances

One of the first things you will do when left alone with liquid nitrogen will be to freeze a short length of rubber tubing and then smash it with a hammer. This is a singularly childish thing to do. It does, however, demonstrate what could happen to your finger if you freeze it in liquid nitrogen. Never wear jewellery when handling liquid nitrogen. If the liquid becomes trapped under it a painful burn will result. Remember that gold is one of the best conductors of heat.

A neglected hazard of liquid nitrogen is the danger of asphyxiation when large quantities of the cryogenic are boiled off in small rooms. Never transport liquid nitrogen in closed vehicles or passenger lifts. Liquid nitrogen can also condense liquid oxygen from the surrounding air. It is therefore bad practice to use liquid nitrogen in contact with organic materials lest the liquid oxygen condensed cause their explosive decomposition.

Solid carbon dioxide can also be an asphyxiant. It is commonly used in conjunction with acetone as a freezing bath. This is an unwise practice since acetone has a very low flash point. Propan-2-ol can be used in its place.

1.6 Compressed gases

Cylinders of compressed gases are potential sources of danger. They should be handled with care lest they rupture. One nitrogen cylinder dropped on its end lost the valve from the top when the shock wave sheared it off. The valve travelled through three concrete floors before coming to rest. Cylinders should always be securely fastened to benches in an upright position when in use. They should be locked upright in trolleys for transportation. They should also be protected from sources of heat since Charles' law applies to their contents just as it does to gases at subatmospheric pressure. Never use oil or grease to ease the fitting of a cylinder head or reduction valve. The outlets of cylinders containing flammable gases are fitted with a left hand thread. However, it is better to dedicate cylinder fittings to a single gas rather than a type of gas. No copper or brass fitting should come into contact with ethyne (acetylene).

1.7 Electrical hazards

When you have finished using a piece of electrical apparatus, unplug it, unless the power supply to the apparatus must be maintained at all times. Removing the plug from the socket prevents a heater being switched on inadvertently or more delicate apparatus being needlessly damaged by a power surge. It also conserves energy and reduces fire risks by cutting of the power supply from the primary coils of transformers.

Since the 1974 Health and Safety at Work Act was not drawn up b scientists, electricity fell outside its scope. This major omission wa rectified by the introduction of the Electricity at Work Regulations o 1989. These regulations, in the form of a European Commission Code replaced much, older legislation. Their main impact on work in a chemica laboratory is their requirement that small, portable electrical appliance (e.g. stirrers, heaters, vacuum pump motors) are tested regularly b competent persons. Given the dangers inherent in operating unsaf electrical equipment in a chemical laboratory their implementation ca only be welcomed. However, there is little point in operating a electrically safe heater if the lead is draped across the hot plate and it insulation allowed to melt.

4.1.8 Radiation hazards

If you are working with radioactively-labelled compounds or in radiatio chemistry, or use X-rays, you will be exposed to the hazards of ionizin radiation. Workers in these areas must seek advice from their university radiation protection officer. Except where the weakest sources ar involved, you will normally have to wear a film badge or monitor the dos you receive. Pregnant students are unwise to work with radiochemicals c radiation sources. The radiation safety code must be strictly observed b students. You should take every care to avoid exposing yourself unnece: sarily to ionizing radiation. The exposure can be minimized by workin under carefully controlled conditions and by paying strict attention t cleanliness and shielding. Always avoid contaminating the immediat working environment and apparatus. Never try to outwit the safet interlocks used to safeguard sources.

4.1.9 New compounds

It is possible that the aim of your project is to synthesize new compound There will be no safety data on these compounds. You must estimate th hazards they will present by extrapolation of the known properties of the congeners. Always consider the worst case and err on the side of safety t being pessimistic. Fortunately very few chemicals are more hazardous tha a consideration of their constituent groups might predict. The same cann be said for mixtures such as gunpowder or impure substances.

You would do well to remember that many of the experiments you fir described in the primary chemical literature are virtually untested. Mar may have been performed only once previously, and could not by ar stretch of the imagination be classed as reliable.

Unanticipated dangers may arise by using reagents from different sources on a different scale under different conditions. Examples include obtaining friction-sensitive precipitates because the laboratory temperature was lower. A product might explode during reflux in an English laboratory at sea level, whereas it did not reach its decomposition temperature in the original experiment carried out at 2000m above sea level in Wyoming.

Further dangers may arise when you try to repeat experiments from the early literature which were carried out in less safety conscious times. You should not, for example, attempt to prepare 2,4-dichloronaphthalene following the original method which uses the banned, potent carcinogen -naphthylamine as a starting material.

When writing your report you have a legal and moral duty to warn others of the possible hazards they might encounter while repeating your work. You should, in turn, gain credit by identifying the hazards that were unrecognized or overlooked in the earlier work you utilized.

.2 CONTROL OF SUBSTANCES HAZARDOUS TO HEALTH REGULATIONS

These regulations came into force on 31 December 1989 and were primarily designed to ensure the well-being of workers in large industrial plants who carry out the same large scale, routine operations every day using a limited range of well known chemicals. It is usually much more simple to ensure compliance with the regulations in these circumstances than in a research laboratory. In the latter laboratory the workers may be using a wider range of chemicals, some of which may be new or little known, in very much smaller quantities in many different operations.

Before starting any relatively untried experiment you should make a simple assessment of the risks involved by rating the inherent hazards of the material on a rising scale from one to three. You should make a similar assessment of the chance of an accident during the experiment using the chemical. The product of these two factors will lie between one and nine. A product of one is ideal since it represents using a virtually harmless material in a safe operation. Such a rating would be achieved by using a salt bridge in a low voltage conductivity experiment.

If the product from your proposed experiment is nine, do not even bother to seek approval. Start looking for another method of achieving your aims. You could reach this high rating by preparing chlorine without taking any precautions to prevent the concentrated sulphuric acid desiccant being sucked back into the flask where it would react explosively with the MnO_4 present.

You should realize that this risk rating system is very crude since heating distilled water in a sealed ignition tube would only give a risk rating of three. Nevertheless, upon the distilled water reaching its critical temperature of 373°C, the resulting explosion could seriously maim you and your colleagues.

The application of the basic formula

$$Risk = Hazard\ severity \times Chance\ of\ accident$$

may suffice for well-understood industrial processes. Normally it will provide too coarse a measure for the operations in your project. Further the hazard data sheets provided with each compound cannot possibly foresee the myriad uses to which chemicals might be put in a research laboratory.

Accordingly it will be necessary for you to carry out your own Control of Substances Hazardous to Health (COSHH) assessment for each experiment you perform. The whole purpose of a COSHH assessment is that every stage from the generation or acquisition of the chemical to the safe legal disposal of the chemical or its products is subjected to the closest scrutiny. If specific safety standards cannot be met at every stage of the process then the use of the chemical cannot be sanctioned.

A COSHH assessment can be made in a systematic manner by following the flow chart shown in Figure 4.1. Using this flow chart will help you to complete the university's COSHH assessment form.

The most obvious initial question to ask about any procedure is 'Do I need to do it at all?'. For example, do you really need to use the benzidine rearrangement as an example of a sigmatropic rearrangement, when many other sigmatropic rearrangements do not yield a carcinogenic product? The next is 'Do I need to do it that way?'. Why make benzonitrile by treating benzene with bromine cyanide or mercury fulminate when it can be made much less hazardously by allowing benzene diazonium chloride to react with CuCN in alkaline solution?

Having satisfied yourself that you have selected the safest reaction, you must now query the suitability of the laboratory. In the simplest case this comes down to deciding whether to carry out the reaction in the fume cupboard rather than on the open bench. More commonly it means you should investigate other aspects of laboratory safety. For example, how might a nitrogen cylinder be secured safely in the gas chromatography suite so that you can use it as a carrier gas? In more extreme cases, it would be your responsibility to determine if the work should be performed in a laboratory equipped with a safety shower. This situation might arise if you wished to carry out large scale nitration reactions. However, any chemical experiment that calls into play such safety apparatus as showers or protective shields is inherently unsafe and should be re-evaluated.

In the next step of the assessment you must consider your own

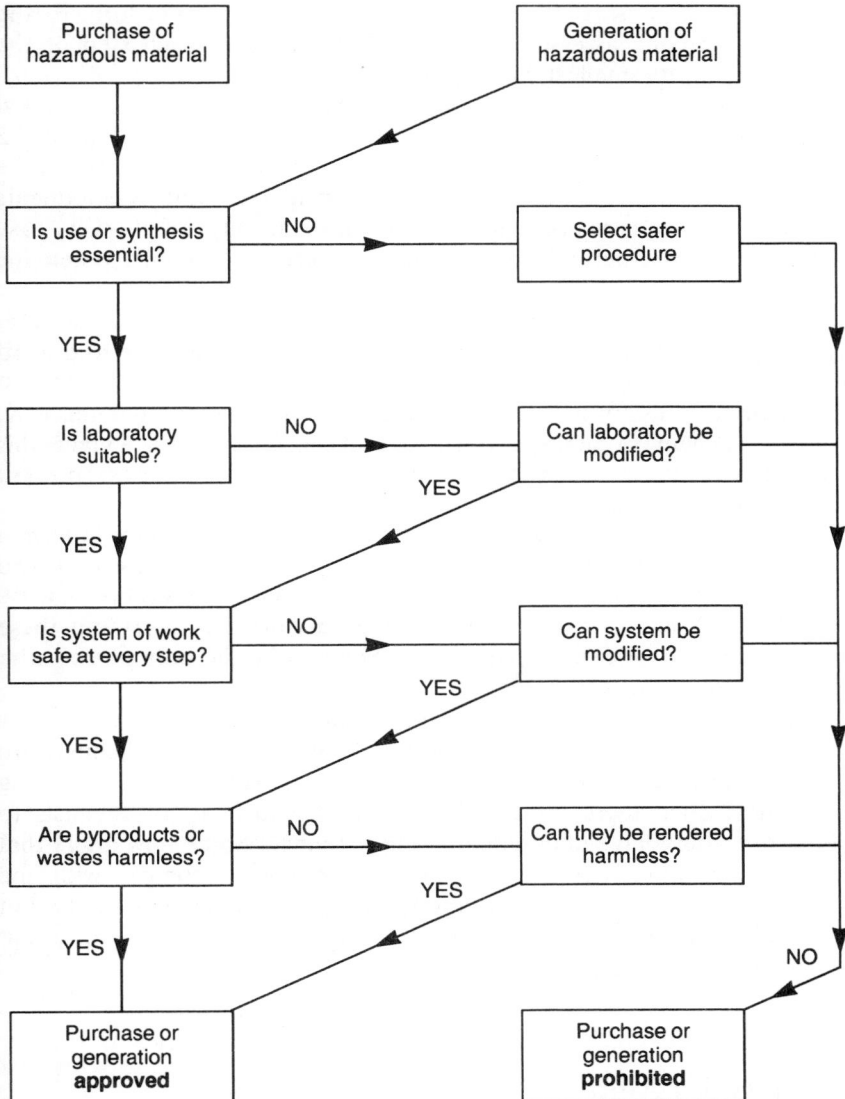

Figure 4.1 Flow chart for COSHH assessment.

knowledge and experience. Have you fully investigated all the risks of the operation? Have you ever carried out a similar operation before? Do you require your supervisor's advice before and guidance during the operation?

The degree of detailed supervision required can perhaps be better estimated once you have carried out the next step of the assessment. This requires you to consider the safety of each step of the procedure. Errors can creep into the assessment by not breaking down each step of the procedure into its smallest components.

As an example, consider the simple, alkaline hydrolysis of an alkyl halide under reflux. The complex operations tree shown in Figure 4.2 arises from this apparently simple process. Most people are surprised at the number of components which need to be considered. Many components are self evident and are taken into account automatically, but it is very easy to make omissions because the complex tree structure inhibits systematic scrutiny.

Another stage that is often omitted in COSHH assessments is the safe, legal disposal of residues from reactions. While it is permissible to discard the 0.001 M KCl calibrant solution for conductivity cells down the nearest sink, neither solids nor most organic materials can be discarded in this way. Consult the chemical catalogues and disposal charts for advice. Even at this late stage you may need to change your experiment should disposal present insuperable problems.

To carry out a COSHH assessment thoroughly it is necessary to have a thorough knowledge of the hazards presented by the chemicals used, and to make a comprehensive analysis of the way in which the reaction is to be carried out. You will realize that you are best fitted to carry out these tasks, since, in a research environment, you will be the only person who knows exactly what is to be done.

Twenty years ago an eminent German chemist, when asked what safety apparatus was present in his laboratory, replied 'German chemists are men!'. He obviously equated scars from chemical explosions with duelling scars. Since then there is little doubt that the attitudes of chemists to laboratory safety have undergone a radical change. Some now argue that safety is becoming obscured by people striving to comply with the increasing quantity of safety legislation and not considering real safety. Let your aim be to know a little about the legal framework and a lot about laboratory safety.

FURTHER READING

1. *Croner's Substances Hazardous to Health*, Croner Publications, Kingston-upon-Thames, UK, 1992 and supplements.
2. IUPAC/ILO/UNESCO *Chemical Safety Matters*, Cambridge University Press, Cambridge, 1992.
3. S.G. Luxon (ed), *Hazards in the Chemical Laboratory*, 5th edn, Royal Society of Chemistry, London, 1992.
4. L. Bretherick, *Handbook of Reactive Chemical Hazards*, 3rd edn, Butterworths, London, 1985.

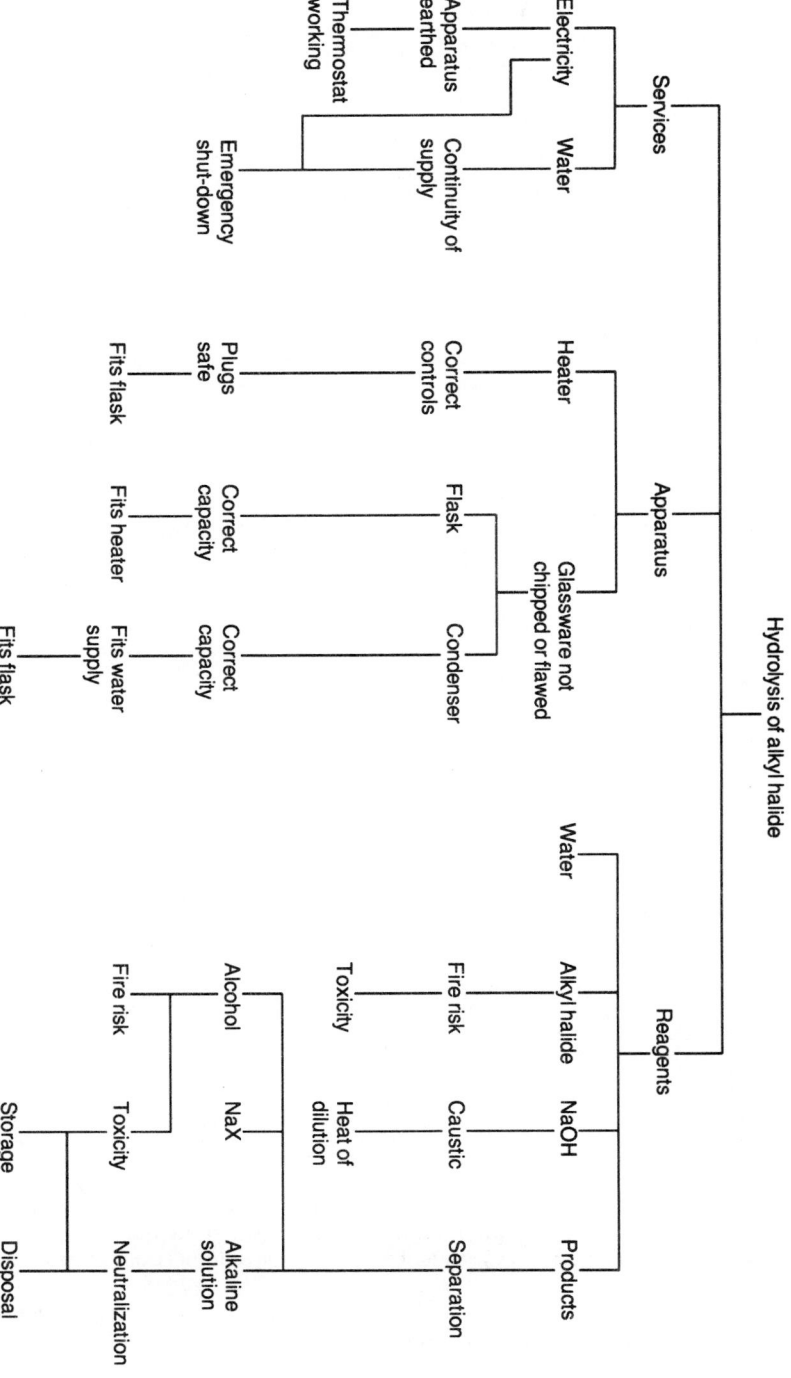

Figure 4.2 Ramifications of a COSHH assessment.

5. J.W. Stranks, *Handbook of Health and Safety Practice*, 2nd edn, Pitman, London, 1991.

In a more modern idiom the Aldrich Chemical Company, The Old Brickyard, New Road, Gillingham, Dorset, SP8 4BR, offer more than 58 000 safety data sheets on CD-ROM. (These are very expensive!)

Carrying out the project | 5

This chapter deals with both the tactical and strategic decisions you will have to make while carrying out the experimental part of your project. It aims to show you how to make the best use of your time and how to meet the departmental requirements during your work. It also includes a section dealing with the possible problems you may encounter during your project work. Advice is given on how many of the most common problems can not only be surmounted but even turned to your advantage.

5.1 RESEARCH STRATEGY

What is the best strategy to adopt to achieve a successful conclusion to your project work? You must first define what you regard as a successful conclusion. Is this the preparation of a new compound, determining the rate of a reaction, or demonstrating the utility of a new analytical method? Are you sufficiently critical of your aim? Should you define it more precisely? Should the first of the above examples be the preparation of a new compound in high yield and purity followed by its unambiguous characterization? Is the second really the determination of the reaction rate to a high degree of accuracy? Should the third also be widened to include high accuracy and low interference from other substances that might be present in the analyte?

Inevitably, the strategy you adopt will be determined to a large degree by the criteria you use to determine success. That different projects will require different strategies is obvious. However, one common feature of all strategies should be to work to the highest possible standards. Just because you are engaged in a preparative project does not mean that you should not measure quantities sufficiently accurately. If you work largely by inspiration (as all the best cooks are reputed to do) you will not be able to report your work accurately enough for others to repeat it.

All strategies should aim to eliminate those errors which arise from unconsidered factors. This requires you to give some thought to what is actually happening in the reaction you are looking at. Is the evolution of gas, by which you are measuring the rate of reaction, solely due to the expected product? Is there some other process going on that produces a large volume of gas, which, even if it only happens to a very limited extent, will make your results erroneous?

Are you introducing some errors by your system of measurement? Adding indicator to an acid/base titration may introduce an error greater than those inherent in using a pH meter in the titration. Are you working in a fashion that minimizes the contribution of systematic errors? Have you prepared two batches of catalyst to see if their efficiency differs?

Procrastination destroys most strategies. It can arise from a reluctance to commence what you regard as a difficult task. This behaviour is also apparent in the animal kingdom. Many birds, when confronted by a rival of the same species will preen themselves or indulge in useless tasks such as picking up stones instead of confronting the rival. Do not imitate the bird. Tasks are accomplished by tackling them. Unattempted tasks are never accomplished. This is not to say that difficult or unpleasant tasks should not be preceded by careful planning, but do not let the problems uncovered seriously delay your start.

Military strategies avoid long advances on a limited front since it is easy for the enemy to cut off these spearheads. Given infinite time and resources a commander would always choose a steady advance on all fronts. Since these conditions are never encountered in real life, practical strategies are a combination of small broad advances combined with narrow thrusts.

It is easy to draw analogies between military situations and those you will encounter during your project. The narrow thrust corresponds to a long, linear, step by step synthesis. The broad approach is suited to finding the optimum operating conditions for a large scale reaction. Physical chemistry projects which depend upon obtaining results after a long period spent learning how to operate a complex instrument could be likened to commando raids. These either achieve spectacular success or abject failure.

Some of the less interesting projects depend upon you obtaining many individually trivial results to establish a pattern of behaviour. These projects correspond to the continuous methodical advance against weak opposition. They may be unexciting, but success of a sort is always achievable, and you will seldom be at a loss to know what task to tackle next.

In Chapter 2 you were encouraged to avoid long linear syntheses lest an intermediate reaction fail to give a satisfactory yield and prevent you reaching the final product. Not only are they dangerous in this respect, but they are also likely to waste much of your time. You may well spend much unproductive time watching the progress of a slow reaction before proceeding to the next stage. You will have difficulty in using this time effectively, particularly at the beginning of your project.

The synthon approach overcomes many of these difficulties in addition to giving higher yields of the final product. If we consider the rival strategies we have routes to the final product **G** as shown below:

Linear synthesis

$$A \rightarrow B \rightarrow C \rightarrow D \rightarrow E \rightarrow F \rightarrow G$$

Synthon synthesis

The synthon synthesis requires fewer steps, which ought to increase the overall yield even if the individual steps are less productive. Further, **H** and **I** can be prepared at the same time which lessens the overall time required for the synthesis. You should always try to incorporate such a strategy into your syntheses to make the best use of your laboratory time. It may not be possible to achieve the ideal solution shown in the above diagram, but you should be able to introduce a few profitable tributaries into an otherwise linear synthesis.

Once you have selected your project you will find it beneficial to give some thought to how you might best carry out the laboratory work. However, do not set yourself unrealistic targets. If you are unable to achieve overambitious aims you will place yourself under unnecessary pressure. There is also a danger of despondency creeping in, which will not put you in the correct frame of mind to complete the remainder of your work successfully. Remember that above all your strategy should be flexible enough to allow for the exciting uncertainties of research work.

5.2 THE METHODICAL APPROACH

Adopting a methodical approach will prevent you from wasting much valuable time. You will find this self-imposed training of great utility later in life. Your project will give you one of your first opportunities to put it into practice. You may already believe that you are adopting a methodical approach to your work. After all, in the second year did you not always hand in your essays and laboratory reports on time?

This punctuality, however, was not the result of a truly methodical approach since the details of your timetable were laid down by others. You were also carrying out predictable, limited and routine tasks. In your project work you will be involved in a network of interconnected tasks. To accomplish these you will have to devise a flexible timetable for tasks that are unpredictable in both their duration and outcome.

Just as in writing, where the most complex ideas are best expressed in the simplest language, so complex tasks are best accomplished by the most methodical approach. It is essential to keep track of all these tasks if you are making efficient use of your time and carrying out several simultaneously.

This is best achieved by making lists of tasks. You may be engaged in a project where you are preparing many derivatives of carboxylic acids using a new reagent. These derivatives will be used in gas chromatographic separations. While the primary object is to determine their retention times on the chromatography column, you will also need to investigate their properties. You may need to obtain infrared and proton or ^{19}F NMR spectra, melting and boiling points, and elemental analyses. It is sound practice to make a table of all these requirements, so that the omission of any analysis or spectrum can easily be detected.

In this way batches of spectra can be obtained without having to set the instruments up for each individual compound. Perhaps it will be more efficient to determine melting or boiling points individually, so that time is not wasted for the hot apparatus to cool between tests. The important thing is that the properties are determined without inadvertent omissions being made.

Much project time is lost simply looking for things that have been mislaid. You will be working in a new laboratory, and, like most students, you will be spending the year in two homes. Even more unfortunate are those living in multiply occupied accommodation who will have their companions' untidiness to contend with besides their own. The best solution is to have a place for everything and everything in its place. Five minutes spent tidying up at the end of a session can save half an hour of searching later.

This applies equally well to laboratory work and to other aspects of your project. At the end of a laboratory session make sure that all the reagents have been returned to their correct location. This should not be too onerous since, in accordance with safety policy, you should have returned these as soon as you had finished using them.

5.2.1 Writing up results

Pieces of paper are much easier to mislay than bottles of reagents. When you are writing up your project you will need to keep track of references, results, draft manuscripts, instrumental output and diagrams. By their very nature instrumental output and diagrams will be large and loose. They will be difficult to store flat in a domestic environment. It is good practice to store this material rolled up in tubes. At the risk of sounding like a children's television presenter, the centres of kitchen towel or foil rolls make excellent free storage tubes. Keep all the tubes together by binding them with a rubber band. Stop the ingress of dirt and dust by stuffing a tissue in the open ends.

Advice on using reference cards has already been given in Section 3.2. If you do not use reference cards, at least record your references in a notebook. This is less flexible than using reference cards but it is vastly superior to using the backs of envelopes or student union handbills.

Results should always be recorded directly into a stout covered notebook. The entries should be made on numbered, dated pages. Results should never be recorded on loose sheets of paper for making fair copy in the future. Why write everything twice? Unlike undergraduate experiments, there is no 'right' answer. Your colleagues will be working on different projects, so there will be no opportunity to compare notes. The results of your project work, like all research, will be unique and must be recorded directly into your laboratory notebook without delay or editing. Discrepancies may form the basis for future investigations either by you or those who follow you later. These discrepancies are not, as was the case in well behaved undergraduate experiments, a reason for losing marks.

As your work progresses you will begin to write the first drafts of your project report. Keep these drafts together in a file or stout folder. It is quite probable that you will be writing these drafts out of order (see Chapter 6). You will also be making amendments and additions to your drafts as your work progresses. This makes numbering the sheets difficult. It is best to use a one- or two-letter code to denote sections. The letters XP might usefully denote your experimental section. The pages in this section can then be numbered XP-1 and so on. If additional pages need to be inserted at any point then they can be denoted by an additional letter. Page XP-7A would follow page XP-7 of the original. You may wish to scrap a heavily annotated version of the experimental section at a later stage. The pages of the new draft of the experimental section could be numbered in the series A-XP-1 to avoid confusion with its predecessor.

If you are typing your text into a word processor you will encounter similar problems. They could even be greater since the almost identical typescripts may lack the visual clues such as the coffee stains or the corrections in green ball point that distinguished the handwritten material.

Label each draft with a different file number, and do not delete the old file until you are certain that you will never require it again. Place the file number at the start of each draft and clip the pages together in some way. Ideally, store the pages of different draft editions in separate files.

Pessimists of the belt and braces school always use two different disks from different makers to save their text from the word processor. They endeavour to store these disks in two different locations and never transport them together in the same briefcase. They also take care to protect the disks from the stray magnetic fields that can be found around NMR spectrometers and electrically powered transport. Do not rest your word processor disks upon the NMR casing nor travel in the motorized coaches of electric trains when transporting your disks.

Those suffering from an advanced degree of paranoia also keep duplicate records of their results as a safeguard against loss or theft. Some supervisors, who have statistical evidence that the probability of a student being mugged or losing project reports and results is inversely proportional to the project time remaining, will also insist on duplicate copies of results being made. You should certainly endeavour to use duplicate disks, since multiuser systems can wipe or overwrite word processor files.

Although research is unpredictable, past results are not. You have no reason to make future unpredictability an excuse for a disorganized present. One of the more important skills you can learn from carrying out project work is the organizational aptitude required to keep track of everything. Another important attribute you can gain from project work is the ability to be always doing something productive that advances your project along the road to success.

5.3 TREATMENT OF RESULTS

In certain projects you may well measure the same property many times. This is particularly likely if you are working in analytical, physical or radiochemistry. When considering these repeated determinations you will need to apply simple statistical techniques to establish the most probable value for the property and the level of accuracy you may assign to this value.

The parameter we use to assign a level of accuracy is the standard deviation. It can be calculated by taking the sum of the squared deviations for each member of the set from the set's mean and dividing it by one less than the number of independent observations comprising the set. Once this fraction has been calculated its square root is said to be the standard deviation. Many sets of observations cluster more or less symmetrically around the mean. These are often described as Gaussian distributions. However, skewed distributions are also known. Perhaps the one most familiar to chemists is the Boltzman distribution of gas molecule velocities. This is heavily weighted towards the slower molecules. Hence before applying statistics to a set of experimental results it is desirable to examine the distribution itself. The concept of standard deviation cannot be applied satisfactorily to a skewed distribution. If the distribution is strongly asymmetrical, it will have to be manipulated first; a log transformation will usually do the trick.

The majority of measurements you will encounter in your project are of a different type. They are results from the interaction of two variables. These do not cluster around a central value but along a line. The pattern of these results often approximates to a straight line, frequently with an intercept upon the y-axis. The line obeys the general equation

$$y = mx + c$$

You should always plot your results to ascertain that the relationship between the variables is indeed linear. Look along the line of scatter to detect curvature or to see if the results at one end of the range exhibit increased scatter. Both these attributes imply some mismatch between your results and the hypothesis being applied. This discrepancy is good raw material for your discussion section.

Assuming that your results are well behaved and show a genuinely random scatter about a straight line you must then decide what is the best line to draw through these points. One school of thought contends that any 'reasonable' line through them will do since the individual results are in error. Since opinions differ as to what might constitute a reasonable line you should draw the line in accordance with more reproducible criteria. The most widely accepted way of drawing a straight line through a set of points (each subject to random error) is to minimize the sum of the squares of the distances of the points from the line. This process, known as linear regression, is analogous to the standard deviation calculated in the one-dimensional case above where the observations cluster around a single value.

An example of the best straight line through a set of points is shown in Figure 5.1. In this case it is assumed that the values of the variable x are known more accurately than the values of the dependent variable y. Thus the coordinates of the experimental points constitute the series (x_1, y_1), (x_2, y_2), (x_3, y_3), . . ., (x_n, y_n), but the best straight line passes through the points (x_1, y'_1), (x_2, y'_2), (x_3, y'_3), . . ., (x_n, y'_n).

As was the case with the deviations from a single, mean value above, it is possible that the sum of the deviations from the straight line may also be zero, since negative and positive deviations may cancel. As before, squaring the deviations eliminates this problem. The best line is achieved by minimizing the sum of these squared deviations. These tedious

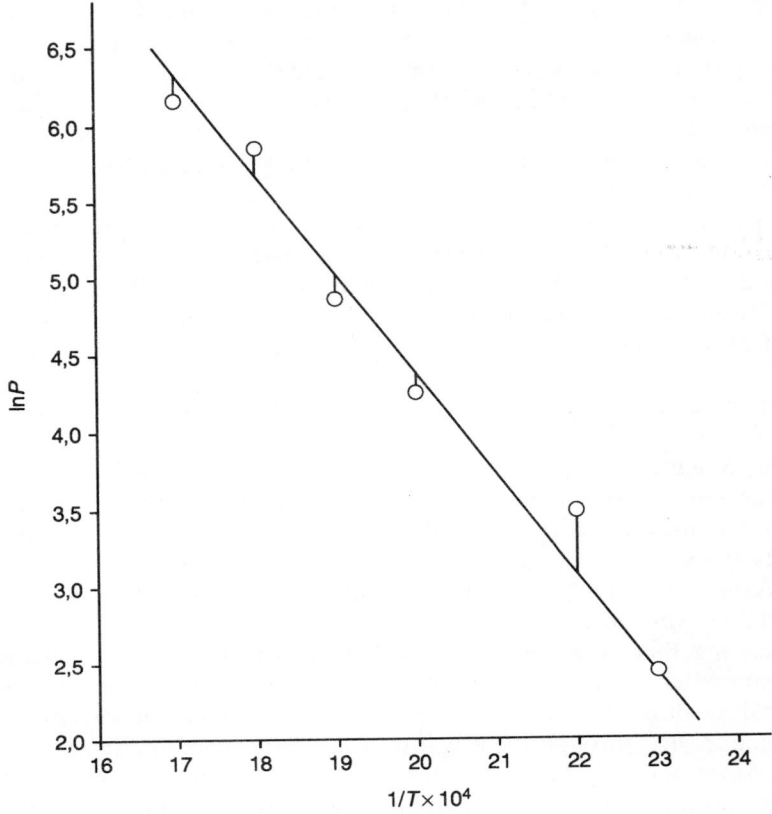

Figure 5.1 Least-squares fit of 'best' straight line.

calculations are best performed by a computer. Most scientific or statistical program packages contain a least-squares program of this type.

The calculations are similar but more involved if the variables x and y are both subject to uncertainty. Full least-squares programs are available for such eventualities.

5.4 WHEN THINGS GO WRONG

It is to be hoped that by intelligent selection of a project, and through your careful initial consideration of the many facets of project work, nothing major will go wrong. However, while few undergraduate projects could truthfully be described as being at the cutting edge of research, many are at least near the edge of the blade. When you are working in virtually uncharted territory it is as well to be prepared for a few surprises. After all, if the results could be predicted with complete accuracy while sitting on a bar stool, there would be no need to exchange the bar counter for the laboratory bench. It is the unpredictability of research that gives it its flavour, excitement and sense of achievement upon completion.

Nevertheless, at some stage in your work it is virtually certain that something, preferably minor, will go wrong. This is why this section is headed 'when' not 'if'. Disasters are seldom as black as you imagine. Even a match-stick looks huge viewed from close to. Inevitably you will be too close to your project to see the problem in its correct perspective. The trick is to turn disaster to your advantage.

Your ability to redirect the project under unfavourable circumstances will be noted, and your success in placing your project back on the rails will redound to your credit. Fortunately the converse is seldom true, and allowances will be made for the difficulties you have encountered, and the honest, if unsuccessful endeavours you have made to revive the project. To draw another military analogy, the true test of generalship is not to lead a sweeping advance against weak opposition, but to conduct an orderly retreat against the odds.

There is always provision in any marking scheme for genuine 'act of God' type disasters. You cannot possibly be blamed for the burst water main outside the department that left your overnight reflux a charred mess in the morning. You can, however, lose marks by bringing about this state of affairs by failing to wire on your condenser tubing. If your negligence wrecks the head of department's office as well, you will probably lose even more marks!

One type of disaster that you might class as an act of God, but which others will not, is being let down by the professionals you may employ to produce various parts of your report. If you employ these assistants then you will be held responsible for their work or lack of it. That is not to say that people will not be sympathetic if your typist is rushed to hospital and cannot complete your report, but you may receive short shrift if your sister wipes your sole word processor disk attempting to print its contents on her office printer after the boss has gone home.

The most common things that may go wrong are a reaction that 'doesn't work', erratic results, a broken or unreliable instrument, or the requirement for greater instrument power, sensitivity or range. If you are following an established procedure you must first make sure you are following it to the letter. For example, one of the author's research students failed several times to prepare a tertiary phosphine because he used anhydrous $CaCl_2$ to dry the product instead of the anhydrous Na_2SO_4 specified in the literature.

If after a second carefully conducted experiment, you are still in difficulty, then you should dissect every stage of the process. Make a detailed examination of each step, however minor or trivial it may seem.

Many problems arise from inhomogeneity. Are you stirring the mixture as thoroughly as the original workers? Have you allowed substance A to dissolve completely before adding reagent B? Have you increased the scale of the reaction and failed to achieve local temperature control or to allow additional reaction time? Is the reaction surface catalysed and have you changed the surface:volume ratio significantly?

Different degrees of reagent purity can also affect the course of a reaction. Have you prepared and pretreated your catalyst correctly? Are your reagents really anhydrous or the conditions anaerobic? Are your reagents of the same grade as used originally?

Erratic results are usually a consequence of poor technique. You should take all possible steps to prevent small errors from creeping in at any stage. Are you conducting your experiments at constant temperature? Laboratories with large, west-facing windows provide excellent examples of the greenhouse effect. Are you making such elementary errors as sometimes holding the pipette bulb in your hot little hand? Are you using matched spectrophotometer cells or merely the first two you found in the drawer? In analytical chemistry projects you must remember to take all possible steps to avoid sampling errors.

Overcoming the above difficulties will seldom merit more than a line or two in your project report. Nevertheless, except in cases of personal error, you should mention briefly the problem encountered and how you overcame it. This helps those who, in their turn, may follow you and use your work as their guide.

Unreliable instruments can bedevil a project. The best solution is to use another instrument. If you can show that the instrument is genuinely at fault then your supervisor should try to provide you with an alternative. This may mean you are allocated time on a research instrument or, less satisfactorily, have your samples run elsewhere as a service. In short, if your supervisor has contacts, do not be afraid to use them. However, you have a duty to your supervisor and your department to submit only samples of the highest quality for these services.

It may be that you can improve the performance of the instrument by altering the operating conditions slightly. If this is the case, it is worth at least a paragraph in the discussion section of your report to illustrate your understanding of the theoretical principles behind the measurement.

One thing you cannot do without your supervisor's express approval is to make alterations to the structure of the instrument in an attempt to improve its performance. Neither should you operate an instrument outside its operating parameters. If you can identify the source of weakness you may be able to have a circuit board replaced, or even substituted, by one of more recent design.

It is more likely that your attempts to extend existing procedures to cognate species will result in failure. The differences in physical properties of congeners may be such that the previous operating conditions cannot be reproduced. For example, trichloromethane is a volatile liquid completely miscible with ethanol. Triiodomethane is a solid with limited solubility in ethanol and attempts to reflux CHI_3 bring about its explosive decomposition at 210 °C.

More subtle differences arise from electronic or steric causes. Copper(II) is a common oxidation state while silver(II) is rare. Solid phosphorus pentachloride is $[PCl_4][PCl_6]$ but phosphorus pentabromide is $[PBr_4]Br$. It is well known that t-butyl compounds frequently have very different conformations and chemical properties from their methyl homologues. Very often projects are devised to detect just such differences.

In these cases negative results are nearly as valuable as experiments in which the two homologues behave in identical fashion. You should not give too much space, nor speculate too wildly, as to the causes of these differences in properties. Always remember that sticking your neck out too far is an invitation to some malcontent to chop off your head.

It was mentioned in Chapter 2 that long, linear syntheses constitute a high risk strategy. A mistake in the seventh stage of an eight-stage synthesis can leave you with no product and feeling mortified. As the synthesis progresses beyond the point of no return it may be desirable for you to run one-third scale pilot syntheses of unfamiliar or speculative reactions. This avoids the chemical equivalent of placing all your eggs in one basket, namely placing all your precious intermediate in one pot. If you adopt this policy, you should adjust the initial scale of your work to allow for material used in pilot syntheses.

Another way of dealing with this difficulty is to carry out a similar synthesis with readily accessible reagents. By doing this you will learn of the pitfalls without wasting valuable and, in the context of your project, irreplaceable material. You should always practice reactions with substances of natural isotopic composition before attempting to investigate isotopically labelled material.

Unless you can pin-point the cause or rectify the situation at the second attempt, you must bring the problem to the attention of your supervisor. You will lose neither face nor marks if you have attempted to discover the root of the problem, your experimental technique is not at fault and you have followed the agreed instructions. This is the point for your supervisor to bring experience to bear on the problem. You can still gain credit from the situation if you can propose solutions of your own. Indeed, your contributions to the discussion can enhance your standing with your supervisor.

So, if something does go wrong, don't panic. Instead investigate the cause and read about the topic to see if anybody else has encountered a similar problem. Armed with this knowledge you may devise, or at least propose, a solution to the problem.

5.5 PERIODIC REPORTS

As Samuel Johnson said 'When a man knows he is to hang in a fortnight it concentrates his mind wonderfully'. Formal, periodic reports to your supervisor or a progress committee fulfil the same function. If the reports are to meet this objective, however, they should not be required too frequently. Twice a term is ideal, while three times a term borders on the excessive. Any greater frequency perverts the project into an administrators' jamboree.

Some students regard periodic reports in the same light as dissidents regard investigation by the secret police. They should not be regarded as scrutiny by Big Brother, nor as a check on your diligence (even if this is the case), but as an opportunity to engage in a considered discussion of the progress you are making in your project. They enable your supervisor to determine your progress and to detect, at an early stage, any of your tangential deviations from the true way.

As with all scientific reports, they should be brief and honest. Early identification of a problem area will give your supervisor more chance to offer help. One of the beauties of chemical research is that there is nearly always an alternative approach that can be explored when the original path appears to peter out. Furthermore, the problem area may give your supervisor a clue as to the future direction the whole project should take. Do not attempt to hide problems. They may be signposts on the way to success.

One of the main benefits you will gain from writing periodic reports will be the discovery of numerous minor omissions that would otherwise have to be dealt with at the last minute. For example, you may find that you have no infrared spectrum of one compound and have never determined the melting point of another. Perhaps you have not had some analysis results returned or you are one point short of completing a graph. If these omissions are left to the last minute you will be unable to deal with them in the most efficient way.

Gathering a set of results together, possibly in the form of a table, may permit you to detect an anomalous result that requires reinvestigation. In other cases collation of results may suggest further experiments that might be used to test the hypotheses arising from your current work.

You should not overlook the opportunities that a periodic report gives **you** to determine the direction your project is taking. The person carrying out the experimental work usually suffers from being unable to tell the wood from the trees. Preparing a periodic report allows you to distance yourself from current problems and thereby obtain a better perspective of the work as a whole.

A further advantage in compiling periodic reports is the writing practice they provide. They will also give your supervisor the opportunity to correct your style before you have written too much in the wrong style. You should find after writing a few reports that you will start to adopt the style your supervisor requires. On the tactical level, you can ensure that you are conversant with the house style required for tables, references, and artwork.

Despite the many benefits that accrue from periodic reports, they should not be regarded as an end in themselves. They are merely stepping stones along the way to a successful completion of your project. You should never regard them as a waste of time, but equally beware of spending too much time on compiling them. Ideally you should match your input to the likely returns they will bring.

5.6 MEETING PROJECT DEADLINES

Most departments are quite strict about project deadlines. They must be, since every student could improve their project by spending extra time upon it. If some are allowed extra time it is unfair to those students who have complied with the regulations and completed their projects on time. While few departments refuse to consider projects handed in after the deadline, most penalize late submissions with varying degrees of severity.

The most obvious cause of projects not being completed on time is lack of diligence by the student. Those who place more emphasis on their social or sporting lives must expect to turn in meagre project reports. Usually it is such students who cause the genuine requests for extensions by people who have suffered from illness or tragedy to be treated so harshly. You should realize from a very early stage that a good project performance depends upon an honest input from you. Projects are like sewers – what you get out of them depends upon what you put into them.

One vital point driven home by making periodic reports is practice in meeting project deadlines. You will undoubtedly have experienced the rush which occurs in obtaining the last few items for inclusion in one of your periodic reports. Do your best to avoid this as far as possible when completing your project report.

Never, ever find yourself in the position of the author who once stayed up all night completing a large review article. In the morning he took the text into his department, photocopied it, caught the underground to Heathrow, handed the copies to his wife saying 'Post these to the editor', kissed her goodbye, and just caught the plane to his summer sabbatical in America. That is most definitely cutting things too fine.

The most usual reason for cutting things too fine is the laudable attempt to include the maximum quantity of work in your report. It is much more important to know when to draw a line under your work so that the report can be completed without its quality being compromised in the rush.

If you have adopted a methodical approach things should be relatively straightforward. However, making maximum use of your time by having several tasks in progress at any instant does militate against them all being completed. Ideally they should all be completed at the same time. On the rare occasions when this does happen you gain the satisfaction of completing a particularly difficult jigsaw puzzle.

Unfortunately it is more likely that at the last minute you will discover some lengthy task has been omitted. These omissions cause projects to overrun. They must be avoided. Applying a still more methodical approach where tasks are listed and ticked off upon completion should help avoid many omissions (Section 5.2). However, this is no defence against the realization, once you have all the other data to hand, that one more experiment would decide the matter. Perhaps if you had given more thought to your project earlier you could have avoided this last minute, blinding flash of insight.

You should agree with your supervisor when the laboratory work should cease. When reaching this agreement you should bear in mind that you are not the only person with a deadline. Your typist and technical artist will also require their deadlines to be set well in advance of the final project deadline so that any errors in their work can be corrected.

It is very easy for both you and your supervisor to hear the siren's voice crying 'Just one more experiment can transform this project into a real winner', and continue the experimental work to the point of danger. You must always assess the time this additional experiment will add to the project. This is not just the laboratory time but the time it will take to include in the text of your report. It could even mean recasting the reference list at the last moment, besides delaying the completion of tables, graphs and associated artwork.

To summarize, the real last minute rush can be minimized by careful planning and by meeting other internally set deadlines. Just as in written examinations, the good candidate is always in need of more time. Unfortunately in project work the candidate must also fulfil the invigilator's task of calling time in the examination.

FURTHER READING

1. E.J. Corey and Xue-Min Chang, *The Logic of Chemical Synthesis*, Wiley, New York, 1989.
2. R.C. Larock, *Comprehensive Organic Transformations*, VCH Inc., New York, 1989.
3. S. Turner, *The Design of Organic Syntheses*, Elsevier, Amsterdam, 1976.

4. S. Warren, *Designing Organic Syntheses*, Wiley, New York, 1978.
5. S. Warren, *Organic Synthesis: the Disconnection Approach*, Wiley, New York, 1982.
6. D.D. Perrin, W.L.F. Armarego and D.R. Perrin, *Purification of Laboratory Chemicals*, 2nd edn, Pergamon, Oxford, 1980.
7. D.L. Massart, A. Dijkstra and L. Kaufman, *Evaluation and Optimization of Laboratory Methods and Analytical Procedures*, Vol 1, Elsevier, Amsterdam, 1978.
8. J.C. Miller and J.N. Miller, *Statistics for Analytical Chemistry*, 2nd edn, Ellis Horwood, Chichester, 1988.
9. J.K. Taylor, *Quality Assurance of Chemical Measurements*, Lewis, Chelsea, MI, USA, 1987.
10. *Minitab Reference Manual*, Release 7, Minitab Inc., State College, PA, USA, 1989.

6 | Writing your project report

At most universities the largest proportion of your marks for the project will be awarded for your written report. Accordingly you should make every effort from the beginning of your project to produce a top quality report.

Your very first task before writing a single word is to find out the department's requirements for project reports. In particular you must be able to answer the following basic questions.

- What are the maximum and minimum number of words allowed in the report?
- What size of paper is to be used?
- Can outside help be used in producing the report?
- Are handwritten reports permitted?
- Are typewritten reports to be single or double spaced?
- How many copies must be submitted?
- Are there special requirements for diagrams?
- Is any particular structure required?
- Are there any restrictions on section lengths?
- Is the report to be bound? If so what margins are required?
- Is the text to be written on one side of the paper only?
- How are the pages to be numbered?

Generally your complaint will be that insufficient guidance is given rather than that too many restrictions are placed upon you. It may well be that the only information available is the maximum number of words. In the absence of any departmental instructions to the contrary you should follow the advice given in British Standard 4821 (1990) *Writing Reports and Theses*. If the two sources give conflicting advice, then the department's must take precedence. However, it is unlikely that there will be many points of conflict since the departmental requirements probably follow the style laid down by the university. This in turn will probably be derived from the British Standard since its research theses must comply with the Standard so that they can be placed on microfiche in the British Library.

You are strongly advised to consult a copy of the British Standard since it contains much sound advice and will draw your attention to previously unconsidered nuances of manuscript production. If you are typing the report yourself it is essential that you read it as you will not have the services of a trained secretary to correct your inexperienced errors. Knowledge of the British Standard will be very valuable should you go on to write a thesis for a research degree.

Most of these points are purely technical and are dealt with in more detail in the following chapter. They have been mentioned here because complying with them will affect the way you write the report. For example, if a restriction is placed upon the number of diagrams, you will have to decide which should be drawn and which can be replaced by text.

6.1 HIERARCHY OF HEADINGS

Writing a long report can be a daunting task at the outset. To make the task manageable and to bring order out of potential chaos you should begin by dividing the report up into sections. Give each of these sections a temporary or working title. However, as George Orwell might have said, some section headings are more equal than others. Having first jotted down all the section headings you consider relevant you should next arrange them in order and make some of them subheadings. In effect you are writing a contents table (Figure 6.1).

To distinguish between headings of different importance you must adopt a presentational style that allows your readers to differentiate between the types of headings you are using. This can be done in several ways.

Normally you will have three levels of headings in your project report. If you have more than four levels of heading in your report you are certainly dividing it into too many sections.

The most common system in use today assigns a number to each chapter. Subsections of each chapter are denoted by a second number separated

CONTENTS

Figure 6.1 Form a contents table should take.

1 PLATINUM METAL COMPLEXES

This is an example of a number one head. It is fully capitalized and centred on the page. It is used for the most important subdivisions of your work.

1.1 Palladium Complexes

This is a number two head, used as the next lower level of subdivision in your report. It is also centred on the page, but only the main words are initially capitalized; 'by', 'the', 'of' or 'and' are not classed as main words.

1.1.1 Ammine complexes

The number three head shown here is the third level of subdivision in your report. Words are not specially capitalized and the heading is placed flush with the left margin.

1.1.1.1 Diamminedibromopalladium(II). If you use this level of subdivision, you are probably dividing your work into parts that are too small. However, if you do use it, place it in italics at the left margin. Note how the heading follows on directly from the preceding paragraph and how the text runs on from the heading in Roman type.

Figure 6.2 Styles used in hierarchy of headings.

from the chapter number by a period. Thus the ten subsections of Chapter 8 would be numbered from 8.1 to 8.10. Each of these subsections may then be further divided. Subsection 8.3 may have only two parts, namely 8.3.1 and 8.3.2. By analogy the next level of subdivision would be given numbers in the style 8.3.2.1 and so on.

It is usual to reinforce the numerical clues to the level of subdivision by using a different style to denote each level as shown in Figure 6.2. You will also note that this is the system used in this book.

An older system achieved the same ends by denoting the principal divisions with Roman numerals and their subdivisions by capital letters. The next two stages of subdivision used arabic numerals and followed by lower case letters. Again each level of heading was associated with a different style and position. The fourth stage of subdivision shown above would be represented by VIII.C.2(a).

6.2 PRINCIPAL SECTIONS

Essentially your report will consist of three principal divisions. These will be the introduction, the results and discussion section and the experimental section. Although all should be written in a common style, each section has its own purpose and its own conventions of presentation.

6.2.1 The introduction

This should set the scene for your readers. It should summarize all that was known about the subject until you came along and confused everybody with your results.

Good introductions are very difficult to write. Paradoxically it is usually better to write your introduction last. By then you will be sure exactly what you wish to include, and the precise emphasis you wish to place on each aspect. By writing it last, you can also take into account any late changes in your experimental programme and make your introduction really fit your topic. Most importantly, you will have become used to writing a scientific report by completing the simpler sections.

Many introductions adopt the historical approach. When this is well done it is difficult to fault. If you have selected a project of which you already have some knowledge, you will be better placed to write a good historical survey. If not, informal discussions with your supervisor, particularly an elderly supervisor, can give you a better sense of the historical background to your project.

You will find that not every topic has developed in the chronologically simple, logical way best suited to a historical treatment. Many topics have advanced in a series of fits and starts, and big advances have often resulted from advances in seemingly unrelated areas of chemistry, or even in other disciplines. For example, the development of flash photolysis as a means of studying fast reactions owed much to Lord Porter's experience in radar in World War II. Similarly the present ready availability of lanthanoid compounds has been brought about by the lanthanoid elements being important fission products of uranium. Thus atomic energy has influenced research in superconductors, phosphors and lanthanide shift reagents.

A more complex pattern of development could make the historical approach unsuitable. This complexity allows you to exercise your scholarly skills by showing how these disparate threads might be woven together to make a coherent theme.

Your introduction must always set your project work in context. You should review the work that has been carried out previously and explain, fully and carefully, how it has influenced your approach. Why are you following, or seeking to extend, an established approach? What are the shortcomings of previous work that have caused you to diverge from established pathways? What other approaches might you have adopted? What advantages will follow from your chosen approach?

6.2.2 Results and discussion section

In this you present your results and comment upon them. It is usually more simple for physical chemists to carry out this task, since they can present their results succinctly in tabular or graphical form. They are then able to comment on them in the text. In these cases you should always discuss the accuracy of your results. A comparison should always be made with any previous results in this or a related area. You should try to comment on any differences, similarities or trends that your results reveal.

If you include a wide range of other work, it is easy to lose track of the points you are trying to make. Looking ahead to any talk you may have to give on your project, there is the added danger of confusing the non-specialist with this excessive detail. The secret is to include just sufficient of the best examples to make your point.

Different problems are presented when there is too little previous work with which to make significant comparisons. It is very easy for you to read too much into your results in the absence of any clues from other work. Try not to paint yourself into a corner by making bold, sweeping statements that could prove too controversial. Always propose a further experiment that could be used to check any assertions you might make. You can always favour one explanation over another, but don't be too dogmatic.

Nevertheless, speculating in this way does give you a chance to exercise your intellect and judgement. Carried out clearly and succinctly it is a very good way of adding weight to this section.

Yet other problems occur in purely preparative work in either organic or inorganic chemistry. You may think that it is difficult to discuss preparations except by briefly stating the yields obtained and comparing the melting point of your product with that given in the literature. This is not the case. Why did you select this method of preparation? How do your yields compare with those of cognate or identical reactions? Why do they differ? Are there any differences (e.g. colour changes, timings) between what happened in your preparation and that recorded in the literature? In cognate preparations did you take into account any special factors (e.g. electronic or steric) and, if so, what changes did you make and why? Were these changes successful or necessary? What strategy did you adopt with regard to protecting groups? Would any other have led to greater yields?

No doubt you will be able to think of other questions that are more specific to your project.

The melting point of your product is a very crude guide to its purity or even identity. What further steps did you take to confirm the identity of your product? Did you record any spectra? Do these spectra agree with the values shown in the literature?

If you believe you have made a new compound, does its elemental analysis agree with the required values? What steps have you taken to determine the structure of your product? What structural evidence have you gained from spectroscopic studies? Is this evidence unambiguous? What further investigations might be made to distinguish between possible structures? If your product is chiral, what is its optical rotation? All these features and more may be worthy of discussion and comment.

Answering these multitudinous questions should give substance to your discussion section. Do not be afraid that many seem trivial to you. This is because you have learned the answers to them while you have been working on your project. Would you have known about the question (let alone the answer) before you started work on the project? Remember that you are not writing your report for yourself, nor are you writing it exclusively for your supervisor. Many of the questions may be self-evident to both you and your supervisor but they are probably far from obvious to people outside the immediate subject area.

It is because of this last fault, failing to state the obvious, that so many project reports do not give a sense of what has been achieved. At the end of your discussion, do not be afraid to state clearly what you believe you have achieved in your project. This statement makes a fitting end to the section, which otherwise might die away on the breeze and be forgotten.

5.2.3 The experimental section

The experimental section is largely mechanical, and should be written first to give you practice in report writing. It should also be written first since it will be the first section you will be able to write with any certainty. Both the results and introduction sections will depend to a greater or lesser extent upon your future results.

The whole purpose of an experimental section is to tell others what you have done and how you did it. Sufficient detail should be given to enable one of your readers to repeat your work exactly.

These details must include what apparatus was used, even down to such mundane details as to which make and type of melting point apparatus was used. All types of apparatus must be similarly identified by make, model type and number. Obtaining these details will cause you to delve into some becobwebbed corners to find the manufacturer's plate on the least accessible part of the instrument.

The sources and grades of the chemicals you used must be stated, together with any purification procedures you may have adopted. This is important when you realize that low levels of an impurity can catalyse a reaction. Low concentrations of peroxides in organic solvents can also destroy many homogeneous transition metal catalysts.

There is no need to repeat at length the preparative details of a reaction you followed exactly. The sentence 'N,N-diethyl-1,4-diaminobenzene was prepared according to directions given in the literature.[107]' can easily be modified to cover most eventualities. On the other hand if you made any alteration, however minor, to a preparation then this must be recorded.

Let your first faltering steps into the world of report writing be guided by the style of presentation adopted by the appropriate publication of the Royal Society of Chemistry. You should have been using this style when writing your laboratory notebook, so the experimental section could be regarded as a less detailed version of your notebook.

6.2.4 Citation of references

You must always acknowledge in your report the contributions of other scientists by indicating to your readers where these earlier workers have originally published their results. By making reference to the work of others you protect yourself from charges of plagiarism, and are absolved from any errors they may have made. References are not cited simply to give a completely unsubtle hint to your examiners that you have read the literature.

There are several ways of citing references in your report, and it is imperative that you find out the exact style of citation required. Changing from one style of citation to another is a certain way of introducing typographical errors into your reference list. Inconsistency of citation styles also annoys your assessors! In the absence of more specific guidance, you should use the house style of the Royal Society of Chemistry. This has the advantage of requiring the minimum number of key strokes. However, it is only suited to the sequential style of citation.

There are three principal systems of arranging references. The Harvard system is seldom encountered in the chemical literature and its use is strongly discouraged. This has the style 'The melting point reported earlier was 170°C (Smith and Jones, 1984).' To obtain full details of the reference an alphabetical list is consulted. This system extends the text and is now falling into disuse as it cannot easily cope with the increasing number of papers being published. If this is the third paper published by Smith and Jones in 1984 to be cited in your report then the citation would have to be amended to '(Smith and Jones, 1984c)'. The main advantage is that the reference list can be revised with minimal disruption to the text.

The second system also arranges the references in alphabetical order

These are then numbered so that Abbott's paper is number 1 and Zymanski's may be number 347. The disadvantage of this system is that the numbers cannot finally be assigned to each reference until the text has been completed. Furthermore, considerable time and effort must be spent checking the inclusion of every reference in the text.

The third system numbers the references in the order they are encountered starting from the beginning of the text. The style of citation for both the second and third systems is 'The melting point reported earlier was 170°C.[21]' Note how the superscript reference number is placed after the punctuation mark to avoid confusion with exponents. Some publishers adopt the house style 'The melting point reported earlier was 170 °C [21].' to be certain of avoiding possible confusion with exponents.

The author has been obliged to use both the second and third systems at different times when writing large review articles for different publishers. It is his opinion that the third system is probably the better until the number of references exceeds about 120. Late additions to or deletions from the reference list using either system create great confusion and difficulty.

The last part of your report will be a list of references. The exact form this list will take will depend upon the style of reference citation you have been instructed (or chosen) to adopt. There will be four main sources of references for your report. These are journals, books, multiauthor review volumes and patents. Each has a distinctive style of citation.

The primary literature which is to be found in the chemical journals of the world is referred to by naming all the paper's authors and the journal name (in *italics*). The journal name is usually abbreviated in accordance with the *Chemical Abstracts* list. These details are then followed by the year of publication, the volume (in **bold** type) and the number of the first page. Thus:

A. Gillie and J.K. Stille, *J. Am. Chem. Soc.*, 1980, **102**, 4933.

Some journals do not have volume numbers and are shown thus:

D.N. Lawson and G. Wilkinson, *J. Chem. Soc.*, 1965, 1900.

Books have the following style of citation:

P.W. Atkins, *Physical Chemistry*, 3rd edn, Oxford University Press, Oxford, 1986, p. 359.

You should note the inclusion of the publisher's name and location, and the year of publication. All this information can be obtained from the title page of the book. You may need to look at the copyright page of the book to obtain the year of publication, since many publishers seem strangely coy about revealing this information.

The volume number (if applicable) and page number appear at the end of the reference. If you are making several references to such a standard undergraduate text, it may be permissible to omit the page number and make one master reference. This avoids padding out the reference list unnecessarily. In this case specific page references may still be possible using constructions of the type, '. . . using the method given in Reference 15, page 102'.

Multiauthor volumes have a similar style of citation, for example:

> D.S. Matteson, in *The Chemistry of the Metal–Carbon Bond*, (ed. F.R. Hartley), Wiley, New York, 1987, Vol. 4, p. 307.

Note the inclusion of the editor's name, in this case F.R. Hartley.

Patents are cited as follows:

> Y. Nagai and I. Ojima, *Jap. Pat.*, 74 110632 (1974); through *Chem. Abstr.*, 1975, **82**, 156488.

This style of citation is used when you have not read the original patent but have read an abstract in *Chemical Abstracts*. Some patents are assigned anonymously to companies, for example:

> Imperial Chemical Industries, *Dutch Pat.*, 6,602,062, (1967); *Chem. Abstr.*, 1967, **66**, 10556.

In this case you have read the original patent but you quote the reference from *Chemical Abstracts* for the convenience of your readers.

If you encounter work by Chinese authors, then you should include all the author's names. Similarly, include all names (except obvious Christian names) of Hispanic authors: in both these societies the patronymic is not necessarily the last name.

6.3 PRESENTATION

Your report must be well presented to be successful. Even the most important and outstanding discoveries can easily be obscured by poor presentation. Everyone will tell you that your writing should be clear and comprehensible. Usually they will then fail to give you any guidance on how these desirable attributes may be achieved. Possibly they believe that once the objective has been stated this state of affairs will simply come to pass. If this attitude is carried over into the writing of the report it will ensure that it is badly written.

No report should be chaotic, but the mnemonic CHAOS describes the main features of a good report which embodies these characteristics:

- Concise
- Homogeneous
- Avoids ambiguity
- Organized
- Simple.

All reports should be concise. A homogeneous report written in the same style with a consistent theme creates a powerful impression. Unambiguous statements increase the power of the report, but without organization the power can be dissipated. Too many reports fail to achieve their objectives because the concepts they wish to convey are not stated in simple language. You should endeavour to embody these qualities not only in the report as a whole, but in every sentence you write.

You have already made major decisions about how your report will be organized, by dividing it up into chapters or sections as discussed in Sections 6.1 and 6.2. However, the organization of each section and subsection still remains. The organization of the subsection requires a different philosophy to the organization of the report. The latter is based on differences in the treatment and contents. The subsection is organized to ensure continuity.

Even a sentence needs organization to avoid circumlocution and ambiguity. Writing concise, short sentences is the best defence against both these evils. If you also remember the acronym KISS (keep it simple, stupid) you will adopt the correct philosophy from the outset. However, if your report begins to read as if it had been written by Ernest Hemingway, you are overdoing the short sentences. Most project reports consist of a pleasant mixture of long and short sentences. This is because most students' normal tendency to write long, rambling sentences overrides the most valuable advice – write short sentences.

6.4 CONVENTIONS

The scientific literature is written in the third person imperfect passive. By being written impersonally it is possible to criticize it impersonally. From your point of view, it means that you should write 'The rate of the reaction was also investigated at 313 K' instead of 'I also investigated the rate of the reaction at 313 K'. Even the first person style is preferable to the imperative 'Investigate the reaction at 313 K'. This last style is best reserved for laboratory instruction sheets. It will soon become automatic for you to write in the third person since you will have become conditioned to it by reading the chemical literature during your literature search.

Many students are seduced by what a postgraduate student called the 'spurious authenticity of the printed page' when he saw his first article in print. Do not allow this view, which has its origins in your own lack of self-confidence, to cause you to copy out large chunks of others' work. Each author has a different style. Juxtaposition of blocks of text copied from two different authors will be just as obvious to a discerning reader as if you had stuck two photocopies of their work next to each other. Minor plagiarism of this kind should be avoided at all costs since it demeans your work. It prevents your own ability from showing through, and examiners may not be inclined to believe you can write originally.

However, if you read an article, book or even a sentence which conveys its meaning well, analyse the reasons for its success. Then try and incorporate these features in your own writing. Do not expect to write in a good and effective style without considerable practice. Practice your style by roughing out drafts whilst 'baby sitting' reactions.

If you are producing your report yourself, do not type your text directly into your word processor. It will not save time. You may be able to type faster than you can write, but you cannot delete or move sections nor make revisions as rapidly on a word processor. Worst of all, you can only see one screen at a time. Looking at 150 word blocks of your text is not conducive to writing a homogeneous report.

It is much better to write out your first draft in long hand in pencil. Write on alternate lines of lined paper or leave plenty of space between the lines to accommodate future alterations or corrections. Read through the text once as soon as you have finished the writing session, and correct any errors you may find. Read it through again the next day to see if it still conveys the correct meaning to you. It is surprising how often you will find that an ambiguity or *non sequitur* has crept in unnoticed while you were in the grip of the literary muse. You will nearly always be able to remove words or phrases. These excisions will almost certainly improve or clarify the text.

On a larger scale, you may find that you have omitted an important point or that your initial choice of ordering the material could have been better. Your first draft should now be corrected, in pen, by making these alterations. If time permits, return to your draft a week later and read it once again. Use a differently coloured pen this time as you may wish to make alterations to alterations.

After making all these amendments you will find that you are now the only person on earth who can read this first draft. By definition first drafts are not suitable copy for typists. It may be possible, however, for you to type a second draft from this mess yourself. A fair copy **must** be made for your typist, and this is advisable even if you are typing your own report. Making this copy provides you with an extra opportunity to improve the text.

It is frequently possible to rescue the pristine sections of your first draft and staple these into the appropriate positions in the second draft to save time. This is known as the scissors and paste method, but staples are faster, more secure and less messy than using paste. Transparent adhesive tape can also be used. Fold the adhesive tape in half and stick it under the added passage, since shiny-surfaced tape cannot be photocopied successfully.

You should make sure your writing is accurate and unambiguous. This is particularly important in scientific work since an inaccurate or ambiguous statement can result in somebody being maimed or even killed. While a recent *Times* obituary stating 'In her late seventies she fed the old and dying with the nuns of Nazareth House' was merely amusing, the vague instruction 'Water and concentrated sulphuric acid were mixed' is downright dangerous. It might be construed as an invitation to add water to concentrated sulphuric acid. How much better it would have been to write 'The concentrated sulphuric acid was added carefully to water'.

There is the additional possibility of making scientific mistakes when writing about scientific subjects, but ordinary grammatical mistakes should also be avoided. Some of the more common mistakes and suggestions for their avoidance are given in Table 6.1.

Table 6.1 Common grammatical and stylistic errors

Incorrect or not recommended	Replace by
Ambiguous abbreviations	
Phosphine (PH_3)	Tertiary phosphine (PR_3)
Non-standard abbreviations	
The bis complexes	The bis(tributylphosphine) complexes
Clichés	
Reinventing the wheel	
Vicious circle	
It is well known that . . .	All these should be
Hopefully	avoided
Meaningful	
Imprecision	
The degree of purity affects the melting point	Less pure samples melt at lower temperatures
Malapropisms	
The Aufbau principal	The Aufbau principle
The principle component was benzene	The principal component was benzene
The Compton affect	The Compton effect
Hydrolysis effects the purity	Hydrolysis affects the purity
It's infrared spectrum	Its infrared spectrum
The isotopes ^3H, ^{14}C and ^{60}Co . . .	The nuclides ^3H, ^{14}C and ^{60}Co . . .
The lightest element, e.g. hydrogen, . . .	The lightest element, i.e. hydrogen, . . .

Table 6.1 *Continued*

Incorrect or not recommended	Replace by
An alkane, i.e. hexadecane, will not react . . .	An alkane, e.g. hexadecane, will not react . . .
Nouns of multitude	
A number of . . .	Several . . .
The majority of azides decompose to nitrogen	Most azides decompose to nitrogen
Tautology	
Yellow in colour	Yellow
Evaporated off the solvent	Evaporated the solvent
Manually operated by hand	Manually operated
Verbiage	
It can be demonstrated that $CuSO_4.5H_2O$ owes its blue colour to d–d transitions	Copper (II) sulphate 5-water owes its blue colour to d–d transitions
You may recall that nitrobenzene is a liquid . . .	Nitrobenzene is a liquid . . .
At this point in time	Now
These are the type of compound that frequently explode on heating	These compounds frequently explode when heated
New to this class of compound is . . .	A new addition . . .
Neologisms	
Engage with	Do, take part
Enhanced	Greater, improved
Finalized	Finished, ended
Focused on	Considered
Hopefully	It is hoped
Impact	Result, effect
Implementation	Doing it
Instrumentation	Instrument
Looking to	Expecting
Outcome	Result
Outgoing	Ending
Prioritized	Ranked
Resource	Material etc.
Share	Show, tell
Simplistic	Simple

Producing your project report $\qquad$ 7

This chapter deals with more specialized aspects of writing your report. Unlike romantic novels, most scientific reports do not consist of simple text. It is quite possible that a well presented chemistry project report might contain tables, structures, graphs, display equations, reaction schemes and photographs.

The contributions that these different features can make to your project and the conventions that must be observed when they are included are discussed in more detail below. Advice is also given on how the artwork can be prepared at the least cost to you by using home-made drawing aids.

The inclusion of artwork and tables in your project report will add considerably to the time it will take you to produce it. You may be unwilling to see to every aspect yourself, in which case you will have to engage specialists (e.g. photographers or technical artists) and provide them with instructions and copy in good time. Remember that your fellow students may be pestering the same specialists at the same time. Even if you propose to have much of this work carried out by the specialists, you will save time (and quite possibly money too) by heeding the points raised in this chapter to make the specialists' work as smooth as possible.

If you know of any specialists outside the university circle, they will be under less pressure at report deadline time, but they may be less familiar with the conventions and house styles that should be adopted. You should glean enough knowledge from this chapter to be able to give them the correct instructions.

The most important decision you have to make about the production of the report is whether or not to type it yourself. Some departments require you to produce it yourself, but in most cases you are permitted to have it typed by a professional typist. The basic dilemma is that chemists are usually poor typists and typists are usually poor chemists. It is a moot point if an undergraduate can teach a typist sufficient chemistry more easily than a chemist can be taught to produce a report of reasonable quality.

7.1 TYPING YOUR OWN REPORT

If you are typing the report yourself you must take care to produce a good quality typescript. Obey the rules of punctuation and leave a single space after all punctuation marks except a full stop where two spaces are required. Do not use 'l' for '1' ('I' is even worse) or 'O' for '0' unless the two numbers are absent from your keyboard. If your printer is going to use proportionally spaced characters, inset paragraphs by using the Tabulator or Indent keys. Otherwise inset paragraphs by at least three and not more than five spaces, but be consistent.

Keep a style sheet and record on this the editorial decisions you have made. This will help keep your manuscript and typescript consistent. For example, it is both untidy and irritating to have 'infra-red' on page 17, 'I-R' on page 23, 'i-r' on page 45 and 'infrared' on page 60 before reverting to one of these forms at the next occurrence of the word. Nearly as many forms of 'X-ray' exist, so decide on one style, record this on your style sheet and then use this style throughout your report. It is worthwhile to record a standard form of all abbreviations on your style sheet.

Typing your own report does have advantages: once sections of your report have been approved they can be typed into a word processor. This enables you to spread the typing of your report over a longer period. Do not pretype large passages if you are using a manual typewriter because their hard copy is so difficult to alter. Word processors encourage sloppy typing because mistakes are relatively easy to correct. Your typing will have to be of a much higher standard if you use a manual typewriter. Reports are written on paper, not graven on tablets of dried correcting fluid.

When using a word processor you can save time by using personal abbreviations for commonly encountered long words and then using the global change facility to write them in full at the end of the typing session. Thus, 'hc*' (a combination of characters unlikely to be found elsewhere in the text) could denote '17-hydroxycorticosterone'; whence 'hc*-21-acetate' could also be altered to '17-hydroxycorticosterone-21-acetate'. This technique can also be used to speed up typing the reference list. Use dummy names for commonly encountered journals, for example use 'j01' for 'J. Am. Chem. Soc.,'. Not only do these substitutions improve speed, they also improve accuracy if care is exercised when making the change.

You can also make use of the word processor's copying facility. You can include all the printer control codes found when typing a reference, and copy it enough times to apply to all the references. For example, references are normally inset, have the journal name in italics and have the volume number in bold type. Each reference ends with a full stop. Making the requests '[inset][italics on][italics off][bold on][bold off].' takes a long time and many keystrokes. Multiple copying of the codes and full stop (by making copies of copies) will save you considerable time and effort.

7.2 EMPLOYING A TYPIST

Some universities will require the students to produce their own reports as part of their training. Elsewhere, many students will not have access to the typewriters and word processors necessary for them to produce their own reports. Others may be slow or inaccurate typists. If the regulations specify a typewritten report, then these students will have to employ a typist to type the report for them.

It is important that you and your typist can function as a team. If you try to carry out the task on the basis of a benevolent master and a dutiful slave it will be doomed to failure. A good typist can contribute greatly to the final appearance of your project and smooth the whole production process for you. At the other end of the scale, an incompetent or antagonized typist can seriously impede your efforts to produce a good report.

The first problem is finding and engaging a good typist. There are several sources of typists who could be employed to type your project report. The best source will undoubtedly be the departmental secretaries. They are conversant with the material they will be asked to type, and with the conventions and house style required for the project reports. However, the departmental secretaries are very perceptive and have had the opportunity to observe you for several years. They will be well aware of your foibles and those of your supervisor. Certain students and certain groups may find that the departmental secretaries are strangely unavailable to type their reports.

If the departmental secretaries are unable to type the report themselves, they may be able to put you in touch with a retired departmental secretary who undertakes typing at home. This is a very good second choice for you. The retired staff should still remember the departmental requirements and the correct way to type chemical and scientific terms. They may not have access to the latest office technology, but if they wish to continue typing project reports in the future they will produce the very best work they can.

Next on the list are the typists who advertise on the notice boards in the department or your hall of residence. These typists can be of very variable standard. While they may be able to type an English Literature PhD thesis superbly, they may be completely baffled by a simple chemical formula. Ask more senior students if they can recommend any of them before approaching the typist or agency.

Some students may be tempted to employ one of their parent's secretaries to type their report. This is not a good idea. They may be very good secretaries, but can they type moderately difficult chemical text and set out tables correctly? Additionally they may live far away, impeding your teamwork and making last-minute corrections very difficult. Furthermore, it is unfair to them, as any difficulties you encounter may sour the working relationship between your parent and the secretary.

Similar strictures apply to inviting a relative to type your project report. Post-graduate students will be able to regale you with tales of how marriages have nearly reached the divorce courts when one partner types, or rather mistypes, the other's thesis.

As a last resort you will have to search the classified advertisements in the local newspaper or telephone directory to find a typist. Once again the relative complexity of the chemical manuscript you produce will eliminate many possible typists. If you are very lucky you will stumble upon an excellent typist who will make an excellent job of your project report. Unfortunately it is more likely that it will be a case of the blind leading the blind towards a second rate report.

Regardless of how you find your typist it is essential to engage one in good time. Remember the best typists will usually be snapped up first. Typists, unlike galley slaves, are not chained to their typewriters and are allowed holidays and other free time. If you engage the typist at the start of your project you will be able to specify the approximate size of your project report and reach agreement on deadlines. Such agreements allow the typists to plan their workloads and organize their lives.

It is important that your typist be reliable, since you will be held responsible for the mistakes and shortcomings of those you employ. The best safeguard against unreliability is the recommendation of others. Lone typists are very vulnerable to having their work disrupted by illness. Young, recently retired typists may well become indisposed at any time in the near future.

When engaging your typist ask to see copies of similar work that have been typed in the past. This request may gratuitously insult the departmental secretaries. In this case go to the library and look at examples of their typing. Do not be surprised if, in return, you are asked to produce a sample of your handwriting. If the typist cannot read your handwriting the partnership is doomed from the start.

You have additional reasons for requesting to see similar work. Can the facilities available to your typist produce the specialized text that a chemistry report requires? Can they cope with subscripts, superscripts, bold and italic type? Is the typeface itself suitable? Are Greek letters available or will they have to be improvized? How will your tables appear?

No typist will type a project report for a box of chocolates or a bunch of flowers. This is not to say that these are not suitable gratuities to acknowledge a job well done, but for your typist your report is a strictly commercial transaction. Three systems of payment are in force. In one the typist charges a rate per hour. In the second a rate per page is charged.

possibly with a premium for tables. The last is a flat fee for the project.

The first system always seem to have great rip-off potential. You may engage a typist who types at the author's rate of 10 words per minute who would charge you three times as much as a moderately competent typist typing at 30 words per minute.

A rate per page seems to be the most equitable method. Typists with pressing debts or extravagant tastes may charge the full rate for a page containing five lines of text. Those who wish to be engaged by other students in the future will usually amalgamate short pages when calculating payment. Very few typists enjoy typing tables, even with the power of modern word processing systems at their command. You will frequently be asked to pay a premium (often 50% of the page charge) for tables.

It would seem sensible to take a specimen, completed project report with you when meeting your typist to agree a fee. The report will allow the typist to make a better estimate of the labour involved.

Whatever system of charges you agree to, make sure that you are in agreement about what is included in the charge. Many typists who charge by the page make an additional charge for the time it takes to make **your** alterations and corrections. All typists should correct **their** errors free of charge.

You should also determine at an early stage of your negotiations the charge that will be made for copies. Normally you will be supplied with a working copy, upon which corrections can be made, as the typing progresses. Once this working copy has been agreed to be in its final form, then no further corrections can be made and a master copy is produced.

If multiple copies of your project report are required, you will then need to decide if these are to be produced by the typist or by photocopying. If there are many diagrams, structures and display equations in your report then photocopying may be the best solution irrespective of cost. If you are to pay for the production of photocopies, ask your typist about charges. The typist may have negotiated a favourable rate with a local copy centre.

It will not set your working relationship with your typist off to a bad start a written agreement covering all the above points is drawn up at the outset. Both of you will then know exactly where you stand and it could save much acrimony later. Indeed, many typists actually welcome such an agreement but are hesitant to suggest one themselves lest they be thought to be totally financially orientated.

Once you have engaged a typist keep in touch with them; reassure the typist that you will still require the project report. At all costs do not let your typist think that you have dropped off the edge of the world. You will find that your typist will be pleased to give you advice about constructing tables so that these will fit onto the page. Your typist should also be able to help you with other aspects of producing your report. The advice you receive from this source may save you much fruitless labour and wasted time. An experienced secretary can often transform a poor quality manuscript into a good report. Nevertheless, you cannot rely on finding such a paragon and it is always your responsibility to produce the best possible manuscript.

Not all business partnerships are successful. If things are going badly between you and your typist then the partnership should be dissolved in both your interests. However, you should take care that you do not initiate the breakdown by providing poor quality, late or illegible copy, and then demanding that a perfect typescript be completed at two hours notice. On the other hand, you may have engaged a typist who can neither spell nor operate the word processor's spell-check program. Neither may you have the time to teach an inexperienced typist the intricacies of producing chemical text. In both these cases you may need to seek another source of help in producing your typescript.

Do not waste too much of the typist's time by trivial enquiries. You have not employed the typist to make polite conversation or cups of coffee when you deliver batches of manuscript. Do not neglect the common courtesies of life, but remember that time is money to typists just as it is to millionaires. In particular avoid contacting your typist at work during normal working hours if you are not the typist's sole employer. This could jeopardize the typist's main job.

The teamwork will really start once you start to produce manuscript copy for the typist. Then you will need to agree on the style of presentation and many other minor matters. The greatest help you can give to your typist will be to know exactly what you require, and to make the style and appearance of your manuscript as close as possible to those of the actual report.

7.3 THE TEXT

You should ascertain the style of presentation since this will influence the layout of the typescript. For example, if the report is to be bound, you should leave a margin of at least 30 mm on the left hand side of each page. Do not make the lines of text too long by using a small typeface or leaving a narrow right-hand margin. About 70 characters per line will be found to be suitable. A typical, double spaced page of text will then contain about 250 words.

Double spacing is to be preferred for chemical reports since single spacing causes sub- and superscripts to become entangled with those on adjacent lines. You should also consider whether to improve the appearance of the text by making it fully justified (like the text of this book). Some word processing systems are driven crazy by long chemical words and you may be unable to use this feature. Make some tests before insisting on full justification.

Normally your report will be written on one side of the paper only. If, for economy, your report will be written on both sides of the paper, you must ensure that the larger margin is on the right-hand-side of even-numbered pages. This implies that odd numbered pages will be the right hand pages of the bound report.

If you are employing a typist to type your report it is as well to discuss the style of presentation required. Always ascertain that the technology in use can cope with the special requirements of producing chemical text. It may be a good idea to make a list of your requirements to give to your typist. A list of agreed requirements delivered with the manuscript could also save arguments later. The requirements might include:

- Leave a left-hand margin of . . . mm
- All pages to be right-hand pages/odd numbered pages to be right-hand pages*
- Top, bottom and right-hand margin . . . mm minimum
- Maximum line length 70 characters
- Number each page at bottom centre/top right*
- Fully/left hand* justify
- Double/single* space the text
- Indent paragraphs
- Do not break chemical formulae at ends of lines
- Avoid widow and orphan lines
- Unless longer than one page do not break tables between pages
- Centre equations, but place number in parentheses at right-hand margin
- Leave suitable spaces (or pages) where indicated for art work.
* Delete where inapplicable.

7.3.1 Mathematical material

Système International (SI) units should be used throughout your report. However, while physical chemists should refer to temperatures in Kelvins, lesser mortals are permitted to use degrees Celsius for mere melting points. Always leave a space between the quantity and the unit, but beware of bad line breaks that leave a number hanging at the end of a line. Avoid very large or very small numbers either by qualifying the SI units with multiplying prefices or by using scientific notation. The latter is recommended since it is more simple to give values the desired degree of accuracy.

Abbreviate the unit when associated with a number but write it in full in the text. Thus '. . . yield 1.2 kg . . .' but 'The same method was used to prepare urea on the kilogram scale'.

Whole numbers up to ten are normally written in full. Other numbers are typed out as, for example, 123. You must never begin a sentence with such a number. '123 samples were analysed' is incorrect. Change the sentence around to 'All 123 samples analysed gave satisfactory results' or provide some further information to avoid starting with a number.

Fractions are difficult to type and it is better to convert all fractions to their decimal equivalents before including them in your report. For example, it is generally better to write 'The electron rest mass is 5.5×10^{-4} of the proton mass' or even 'The electron rest mass is 0.00055 of the proton mass' than 'The electron rest mass is 1/1836 of the proton mass'. Even this last form is still better than splitting the fraction on two single spaced lines thus:

line 1 $\quad\quad$ 1

line 2 $\quad\quad$ 1836

which in turn will present many less problems than writing it on a single line thus:

$$\frac{1}{1836}$$

Both the bar fraction representations are inelegant, since the numerator is not centred over the denominator. Additionally they require you to be very conversant with the capabilities of your word processing system and the characteristics of your printer. They will undoubtedly take longer to type than to calculate as decimals and the last form will not print successfully on anything but the most advanced laser printer. It is customary to type decimal fractions with a leading zero as in 0.123 rather than .123. This also applies to table entries.

Any mathematical equations in your report should be numbered. You should use a different numbering system (which also looks different) for these to the one you used for chemical equations.

7.3.2 Fonts

Printed scientific works use different fonts, either through long-standing custom or as a means of clarifying the text. The most commonly used fonts are *italic*, **bold**, superscripts and subscripts.

Representing these correctly in your manuscript or typescript may cause you and your typist some problems. Italics are denoted in both manuscript and typescript by single underlining, and bold type by wavy underlining (Figure 7.1(a)). Most typists are not conversant with proof marking systems used in publishing houses. It is better, for example, to place sub- and superscripts carefully in their correct places in the manuscript than to

The complex cis-[Pd Me₂ (PPh₃)₂] eliminates ethane intramolecularly under, these conditions. The ligand 7 forms a trans-complex which ~~does~~ cannot eliminate ethane on heating.⁴³

(a)

The complex $\underline{cis}$-$\left[\text{PdMe}_2(\text{PPh}_3)_2\right]$ eliminates ethane intramolecularly under these conditions. The ligand 7 forms a $\underline{trans}$-complex which cannot eliminate ethane on heating. [43]

(b)

The complex cis-[PdMe₂(PPh₃)₂] eliminates ethane intramolecularly under these conditions. The ligand 7 forms a $trans$-complex which cannot eliminate ethane on heating. [43]

(c)

The complex *cis-* [PdMe$_2$(PPh$_3$)$_2$] eliminates ethane intramolecularly under these conditions. The ligand **7** forms a *trans*-complex which cannot eliminate ethane on heating. [43]

(d)

Figure 7.1 Appearance of: (a) manuscript; (b) typewritten text; (c) dot matrix printer output; (d) laser printer output.

use a circumflex over subscripts and a small vee underneath superscripts.

If you or your typist are using a simple, manual typewriter, then the best attempt, within the limitations of the machine, may look something like that in Figure 7.1(b). The limited keyboard of the typewriter had no square brackets so these have had to be added by hand. They can be drawn in freehand, but a neater result is achieved by using a stencil and a 0.35 mm drawing pen as shown here. This pen has also been used to underline the italics and bold type in the text. A cheaper solution would be to use a fine, black, ball point pen. The text must be double spaced when using a manual typewriter's full sized numbers or the sub- and superscripts of adjacent lines will become entangled. The super- and subscripts have been obtained by moving the typewriter carriage one notch backwards and forwards, respectively. It is usually quicker to leave spaces for subscripts in long chemical formulae or in equations, and then notch the carriage forward before inserting all the subscripts on the current line.

Laser or dot matrix printers usually handle these problems much better. However, this versatility is achieved at the expense of cluttering up the text with printer control codes. Their inclusion slows down the rate of typing the text since each control code usually requires several key-strokes. Further, fonts not only have to be turned on, they also have to be turned off, which requires still more key-strokes. Failure to turn off a font means the text continues to be printed in the unwanted font until the next change of font is ordered. Time and keystrokes can be saved by incorporating font changes in macros.

One problem that can arise with some dot matrix printers is that they advance the paper by one dot (0.12 mm) each time a subscript or superscript is turned off. A long ionic formula with many sub- and superscripts can cause a line of text to droop alarmingly (Figure 7.1(c)).

Another problem that occurs when you wish to print sub- or superscripts in italics is the order of requesting the fonts. Thus when wishing to print the final character of the point group C_{3v} do you request subscript or italics first? The order will depend upon the font hierarchy adopted by your word processing system or printer. Finding out this information means searching through the badly written, obscure sections of your instruction manuals, so it is usually quicker to carry out a few trials. Some printers merely become confused by the double instruction and print the point group as C_{3v}.

Besides using different fonts, scientific papers also use different alphabets. Greek letters are widely used in chemistry to denote geometrical isomers such as γ-lactam, or stereochemistry {e.g. $\Delta(-)_{589}[\text{Co}\{(\text{R})-(-)\text{pn}\}_3\lambda\lambda\lambda]^{3+}$}. They are also used in spectroscopy (e.g. $\pi \rightarrow \pi^*$), crystallography (e.g. 2θ scan), mechanistic studies (e.g. β-hydride elimination), thermodynamics (e.g. ΔH°) and radiochemistry.

Since many word processing systems and their associated printers cannot cope with Greek letters, it is as well to be prepared to draw them in by hand. A few preliminary tests will enable you to make suitable provision in your manuscript. If you change to a higher quality printer for your final printout, do take care to ensure that it can also cope with unusual characters. Greater cost, speed or print quality do not necessarily imply that a wider range of characters will be available.

If you do have to draw the Greek characters in by hand, ensure that you leave space for their insertion. Word processing systems always ignore spaces at the beginning of a line, so if the first word on a line was 'β-ray' merely leaving a space will mean the new line begins '-ray'. When this happens you will have to break the left hand margin when you insert the character β by hand. To avoid such untidiness you can substitute dummy Roman letters for the Greek characters. You should be able to alter these letters unobtrusively to the required character. For example 'n' can easily be converted to π. Often italic characters are more useful. A single stroke converts 0 to θ, while beta can be drawn by adding an extra loop to p. Similarly, the addition of tails to u or n converts them to μ and η, respectively. Even symbols can be pressed into service; \ is a good starting point for λ, as is _ for Δ.

Always alert your typist to the presence of Greek characters by making a note in the margin of your manuscript. These notes should follow the style 'Gk beta', 'uc Gk delta' and so on. The Greek alphabet can be found in Appendix 1.

Symbols may also have to be drawn in. The most common are long reaction-arrows or the reversible reaction sign. Neither does the London Transport logo superscript (e.g. ΔH°) so beloved of thermodynamicists appear on any keyboard, extended or otherwise.

When typing out mathematical material you should remember that variables should be italicized. However, functions (for example cos, $\log_{10}$, ln or sin) are normally printed in Roman type.

7.4 TABLES

Tables are a powerful way of organizing numerical data. They are capable of greater accuracy than the equivalent graphical presentation (Section 7.5). Furthermore, you can use tables where graphical presentations are impossible, as is the case with tables of analytical data.

Each Table should be numbered sequentially in the text and have a brief, descriptive heading as shown in Figure 7.2. Each column should have

Table 8. Physical properties of complexes containing two tertiary alkyl-phosphine or -arsine ligands

Complex	Conductivity /Siemens[a]	Anion peaks /cm^{-1}[b]	Anion symmetry
$[Cu\{P(C_5H_9)_2Ph\}_2][BF_4]$	28.7	1100	T_d
$[Cu\{P(C_5H_9)_2Ph\}_2][ClO_4]$	28.0	1105	T_d
$[Cu(FBF_3)(PCy_2Ph)_2]$	20.6	1080, 1035	C_{3v}
$[Cu(PCy_2Ph)_2][ClO_4]$	27.0	1100, 600	T_d
$[Cu(OClO_3)\{As(C_5H_9)_3\}_2]$	24.7	1130, 1030	C_{3v}

[a] Determined on mM nitrobenzene solutions.
[b] Nujol mull spectra.

Figure 7.2 Appearance of a typical table.

a heading that shows the property which is the subject of the column. Where appropriate this property should be divided by the unit of measurement so that mere numbers appear in the body of the table. These column headings should be ruled off from the table heading and the body of the table. Do not use a page-wide series of underscores to construct these rules, as the underscores do not make a continuous line. If you draw in the rules by hand, triple spacing the text at the point they occur will give your tables a neater appearance.

Tables should be constructed so that they lie vertically on the page. Vertical tables can often be obtained merely by transposing the columns and rows of a horizontal table. Only in extreme circumstances should horizontal tables be included. If your report is to be preserved on microfiche, your readers will have to lie along the bench to read horizontal tables!

Remember that your standard line length will be about 70 characters. Tables that require more characters per line than this will have to be revised or printed horizontally. Horizontal tables can accommodate a maximum of about 120 characters per line (including spaces).

One cause of excessive table width is the presence of invariant or near invariant columns. The former should never appear in any table. The information can be conveyed in either the title or a footnote. Near invariant columns can be avoided by using footnotes. Suppose you have a series of reaction rates measured by 298 K except for one entry where the temperature was 293 K. Instead of having a column for temperature you could entitle the Table 'Rates of esterification reactions at 298 K' and place a superscript italic letter after the rate obtained at 293 K. In the footnotes to the Table '^{f} at 293 K' might then appear.

Footnotes can also be used to convey additional information whose inclusion would make either the title or column headings too unwieldy. Figure 7.2 shows how footnotes can be used in this way. Because table entries are usually numeric you should normally use letters to denote footnotes. The footnotes appear after the page-wide rule drawn below the last row of the table.

If you are typing a table yourself you will find it needs careful planning to achieve a neat layout. You should leave a minimum of three spaces between columns and have the columns evenly spaced across the page by increasing the intercolumnar spacing equally where this is possible. Numerical entries in a column should be aligned on the decimal point.

Tables are difficult and time-consuming to plan and type. Once you have tried to construct a successful table you will see why typists usually charge a premium to type them. The labour involved may tend to influence the number of tables you wish to include, particularly if you are typing the report yourself.

When drawing up the structure of a table try to ensure that the rows and columns are placed in some kind of logical order. If a column depends upon the data contained in another column place it to the right of the latter. Arrange the entries in the table either in decreasing or increasing order of a key property. If there is no trend amongst the properties then arrange the entries in alphabetical order. Another valid system might be to place your entries in order of increasing relative molar mass for organic compounds or in order of increasing atomic number of inorganic species.

Tables should be placed on a single page. If the table is too long to fit on one page then the second page should have the column headings repeated at the top. These headings should be ruled off as they were at the start of the table.

7.5 GRAPHS

If the trend shown by the data is more important than their precise values, you should consider presenting the data in graphical form. It is not good practice to present the same data in both graphical and tabular form.

Graphs are becoming more popular as many word processing systems can now cope with their inclusion in the text. However, the golden rule is 'If you can tell it was drawn by a computer, it isn't good enough'. That is, a 45° line should be represented as a line and not as a staircase. More or less steeply inclined lines drawn by inferior computer systems show even more obvious breaks of slope. Another feature that mars many computer-drawn graphs is the inappropriate scale of the axes' labels. Further, the range of characters available for these labels is often insufficient for chemical needs.

One sure way of achieving a poor-quality diagram with current technology is to scan a diagram in a journal and then incorporate this in your text. If you want a really bad quality diagram try enlarging your scanned diagram for inclusion in your project report.

You may think that the only alternative to computer generated diagrams is to equip yourself with a drawing board and a full set of draughtsman's instruments and pens. A decade or more ago this was the only way you could have drawn neat diagrams. Fortunately a felicitous combination of both methods can be used to prepare eminently acceptable diagrams.

A laser printer can be used to prepare axis labels and even the scales' numerals. These can be rendered in the correct size using a variable magnification photocopier. The bald body of the graph can be drawn either by a computer and printer giving high resolution graphics or by yourself using a drawing pen and tracing paper. The variable magnification photocopier can be used to reduce the pasted-up original to its final size. Reduction to final size is preferable to enlargement to final size since any minor imperfections are also reduced.

If you are fortunate enough to own a drawing board and tee-square, or can borrow them, then use them when you draw your graphs on tracing paper. If you have no access to them, there is no need to stand in front of a graphics supplier's window calculating how many meals you will have to forego to own them. You can make an adequate drawing surface by obtaining a large, smooth cardboard box from a supermarket and taping several sheets of this cardboard together to give a stiff, smooth surface about 500 × 500 mm. Tape an A3 size piece of graph paper, ruled in millimetre squares, to the top surface to complete your drawing board. The graph paper allows a transparent ruler to replace an expensive tee-square. Drawing board owners should also fix a similar sheet of graph paper to their boards before starting drawing.

For special purposes other types of graph paper can be used as a base. Phase rule investigators may wish to use triangular graph paper. Logarithmic paper may be more suitable for kineticists or radiochemists.

Unfortunately there is no alternative to buying a drawing pen. These are available at many stationers as well as graphic suppliers. Several makes are available, and for the small use you will make of a pen there is no reason not to buy the cheapest. The pens come in many sizes and a nib width of 0.5 mm is recommended. The slightly stepped nibs of stencil pens are to be preferred since they make it more difficult to smudge lines drawn against a straight edge.

Tracing paper should always be used. Not only can you trace the graphs from rough copies but it gives cleaner reproductions than typing paper. The latter, being softer, less smooth and more absorbent than tracing paper, tends to pick up as the drawing pen moves across its surface and gives a feathered edge to any lines drawn.

More pronounced feathered edges can be produced by allowing the ink from the drawing pen to be drawn by capillary action underneath the ruler and related drawing guides. This can be prevented by modifying drawing guides intended for pencils or ball-point pens. If you place small pieces of sticking plaster underneath rulers, set squares, protractors, or other drawing guides that would otherwise lie directly upon the drawing's surface you will prevent the ink from seeping underneath them, even when you incorrectly hold the drawing pen at an angle to the guides. The sticking plaster will also help to prevent the guides from sliding over the surface of your drawing while being used. Take care not to smudge your diagrams by moving drawing implements across new lines that are still wet. Drawing ink takes a long time to dry on tracing paper when the humidity is high.

Another advantage of tracing paper is that guide lines may be made on its surface in pencil before or during drawing with ink. Once the ink is thoroughly dry, the pencil lines can be erased using a soft rubber eraser. It is even possible to erase ink lines that have been misdrawn or smudged on the tracing paper by using a special ink eraser which is available quite cheaply from graphic suppliers. You will find that small mistakes and minor smudges can be removed more conveniently by gently scraping the tracing with the corner of an old razor blade. Smooth over the scratched surface with a pencil eraser before reinking the offending portion. Burnishing with a pencil eraser after using an ink eraser is also beneficial. The pencil eraser can also be used to generally clean up any drawing once the ink has dried thoroughly.

Tracing paper is available in several weights. The $90\,\mathrm{g\,m^{-2}}$ weight is the most suitable. Lighter papers are too flimsy and do not withstand rough usage. Heavier papers are a waste of money. You can protect your drawings by storing and transporting them rolled up in tubes. Do not buy storage tubes from your graphic supplier. The discarded centres of kitchen rolls take A4 diagrams and the centres from 300 mm wide rolls of kitchen foil are a perfect fit for 298 mm wide A3 diagrams.

Axis labels can be prepared not only from computer print out but by the 'poison pen method'. This method consists of cutting out the relevant words and figures from the photocopied text of a chemical journal or catalogue and sticking them on the bald diagram photocopy. If all the above strategies fail, you will have to draw the axis labels using a stencil. If you will also need to draw structural diagrams or display equations (Section 7.6) one of the very specialized chemistry stencils will serve to label axes as well. Advanced chemistry stencils also contain a set of numerals in a smaller font. You will find these invaluable for drawing correctly proportioned sub- and superscripts. Otherwise buy the cheapest 5 mm alphanumeric stencil you can. This should be compatible with your 0.5 mm pen.

Fill your pen with black drawing ink and cap it firmly when not in use to prevent the nib drying and clogging. Try to use if frequently to prevent this. If it should clog, shake it horizontally (away from important work!) to dislodge the blockage. In severe cases soak the nib unit overnight in luke warm water. Never use boiling water or organic solvents on the nib unit.

Straight lines can be drawn against a straight edge. Curved lines can be drawn against a flexible curve or by using french curves. To stencil, firmly fix a long ruler to your board (or use the tee-square, locked in position where possible) to support your stencil. Slide the stencil along this edge and use the pen vertically in its letters to write out the numerals and labels. It is as well to practise using the stencil upon typing paper before using it upon your prize diagrams. You can make this practice pay by drawing the axis labels or display formulae. Save these attempts and measure them so that you can evenly space your final copy on tracing paper.

In all cases try and draw your diagram in the correct proportions to fit the space available. As with tables try to make the graphs fit the vertical rather than the horizontal format. Graphs can be spoilt not only by inappropriately scaled labels but also by inappropriately sized points. A good size to aim for is about 5% of the longest axis, but if many points are drawn they will need to be smaller. Very large points, drawn to encompass

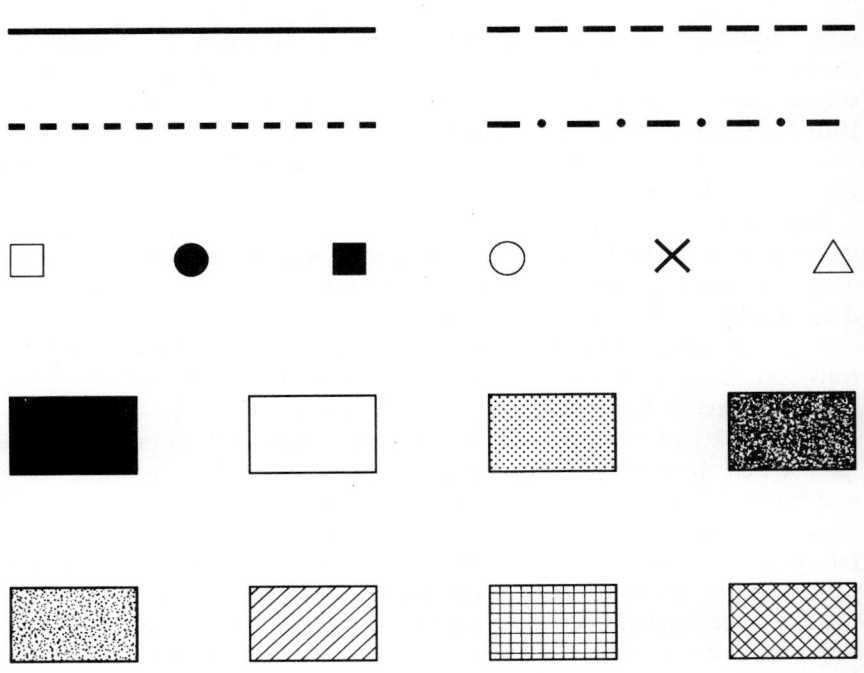

Figure 7.3 Types of lines, points and shading used to distinguish different experiments.

experimental errors, should be avoided. The errors should be indicated by bars extending from the points.

If you are including several curves or lines on one graph these should be clearly distinguished from each other. Different points (Figure 7.3) can be used for each set of results. Invert your stencil and use the vee to draw triangles, or use the sides of capital ens to draw squares, filling in the vacant side(s) in either case using a ruler. Better still (or in addition) label each curve with a capital letter. Use the figure legend to explain which curve is which by referring to the letter.

The master diagram can be prepared from the tracing by photocopying. Remember to place the diagram face-downwards in the photocopier or you will need to issue all your readers with a mirror. Also place a sheet of clean, white paper behind the transparent diagram or you will obtain an interesting photocopy showing the interior of the machine as well. Any labels can be stuck on the master photocopy with an office glue stick. Store the master diagram is a dry place to avoid these labels peeling off.

Some of the numerous ways of presenting graphical information are shown in Figure 7.4. Bar or column charts (Fig. 7.4(a)) are not widely used in chemical publications. If you are including several different columns or

Figure 7.4 Presentation of graphical information: (a) column chart; (b) open graph; (c) boxed graph; (d) boxed graph with grid.

bars upon your chart these can be distinguished by different shadings. To save time leave the largest total area unshaded. The smallest total area can be a full black. Very large full black areas can cause problems for photocopiers having a bad day. The stippled and scribbled areas shown in

Figure 7.3 can be drawn quickly without recourse to expensive apparatus. Successful, regular cross-hatching is difficult to achieve even when using top quality technical drawing equipment and should not be attempted when using a cardboard drawing board. Another advantage of stippled or scribbled shading is that they cannot cause optical illusions which may make the bars or columns appear curved.

Line graphs (Figure 7.4(b–d)) are more common and differ mainly in the way the axes are shown. The simplest merely has two axes and is known as the 'open' type. The next stage is to complete the axes around the remaining two sides to form the 'boxed' version. The final stage of development is to include the scale grid within the box. This type, although common in the *Financial Times*, is not recommended for your project report. Like the box type it requires additional lines to be drawn. It requires a second pen with a nib half the width of that used to draw the remainder of the graph to draw the scale lines within the box. Further, upon reduction, the thinner lines are often either lost or compete with the more important lines. If the exact values of the points are so important as to require an internal scale then the information should be presented in tabular form.

7.6 REACTION SCHEMES AND OTHER FIGURES

You can generate inorganic reaction schemes in a similar way to graphs by using a laser printer and the poison pen method. Reaction schemes show the conversion of starting material A to product B by several steps or routes, or show the sequence of steps in a mechanistic pathway. By tradition the starting material, isolable intermediates and products are shown 1.4 times larger than the reagents which appear above the arrows. All that is required is to make an additional 70% reduction of those reagents and conditions that will appear above or below the arrows. The reaction schemes should be numbered sequentially, i.e. 'Scheme 1', 'Scheme 2' and so on, throughout your report. It is usual to give each scheme or figure a concise, descriptive caption. Brevity can be achieved by omitting verbiage of the form 'A scheme showing . . .', or 'A histogram of . . .', since it should be clear that the art work is either a scheme or a histogram, respectively.

A similar type of illustration but which is classed as a figure is the 'Reactions of . . . ' diagram. An example, prepared by the poison pen method from laser printer output, is shown in Figure 7.5. In these figures there is no clear pathway as there is in a scheme. Note the economic method of dealing with the reagents normally shown upon the reaction arrows. This uses lower case Roman numerals on the arrows and gives the reagents and conditions used in the figure's caption. The arrows have been

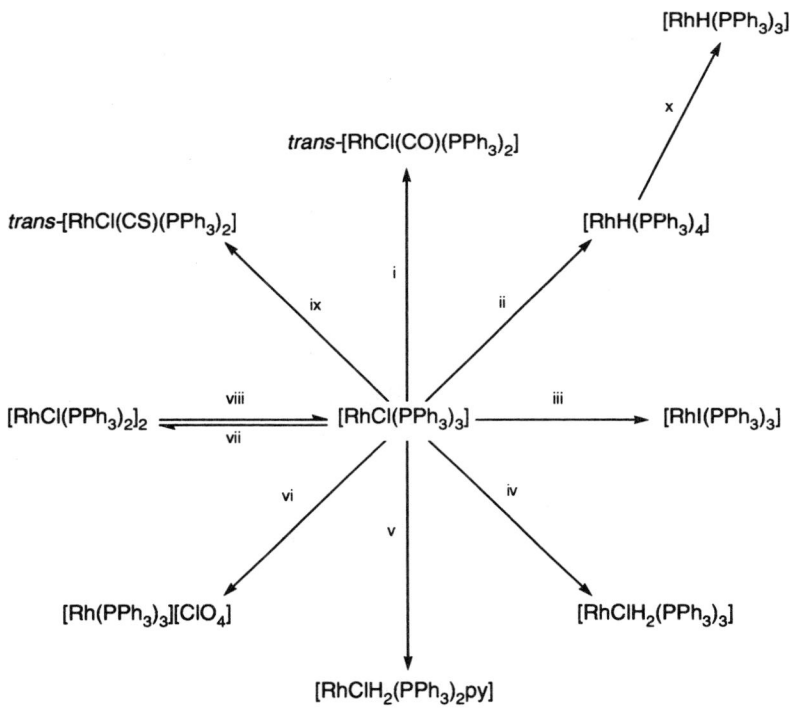

Figure 17. Reactions of $[RhCl(PPh_3)_3]$: (i) RCHO, C_6H_6/EtOH; (ii) $NaBH_4$, PPh_3, EtOH; (iii) $SiMe_3I$, 25°C; (iv) H_2, PPh_3, C_6H_6; (v) H_2, py; (vi)$TlClO_4$, Me_2CO; (vii) C_7H_8, 109°C; (viii) PPh_3, C_7H_8, 25°C; (ix) CS_2, MeOH; (x) -PPh_3, EtOH.

Figure 7.5 A typical figure produced by the 'poison pen' method.

drawn in with a drawing pen; solid heads to the reaction arrows disguise small imperfections when they are drawn by hand. Some stencils have arrow heads included in their symbols. These are difficult to position accurately and filling them in with ink again disguises small placement errors.

Organic schemes and figures present more problems, since you will normally wish to convey structural information. Advanced, specialized, computer graphics programs are beginning to cope with this problem. However, many of the best systems are not yet compatible with all the word processing packages in common use. Lesser packages do not yet produce acceptable results (Section 7.3). Using them will seriously reduce the quality and appearance of your report. The main disadvantage of all

Figure 7.6 (a) Display equation; (b) structure.

computer packages is the length of time it takes to become proficient in their use.

Display equations (Figure 7.6a) can be regarded as small figures. Ensure that when they are reduced they will fit between the margins of the page. Do not be tempted to reduce the length of the arrow to achieve this, as you may wish to include reactants or conditions above and below this. If necessary draw them on two lines, breaking the equation after the arrow. Inset the products from the left hand margin. Ideally all your display equations should be reproduced at the same magnification. This is also the cheapest solution since several originals can be placed in the reducing photocopier at once.

Alternatively, you can use one of the relatively expensive, specialized stencils referred to above. These are usually produced for the 0.5 mm drawing pens, although smaller versions are also available which are compatible with narrower-nibbed pens. If you are hoping for a future career in research or teaching, they may prove to a worthwhile investment (although they may be overtaken by advances in computer graphics).

Possibly your department may have a stencil for student use, or it may be possible for a group of you to buy a communal stencil. It may be the most economic course for you to have a few diagrams drawn professionally. Ask the technical staff of the department or the university's visual aids service if there is anybody who offers this service.

Rub-down lettering may seem an alternative to physically drawing diagrams but this is also an expensive solution. Rub-down sheets of chemical structures are available and these can be supplemented by the more familiar alphanumeric sheets. However, workers in zinc chemistry will run out of zeds long before they run out of ens since the distribution of letters on the sheets reflects their occurrence in written English.

If none of these solutions is appropriate, a little improvisation will enable you to produce adequate structural formulae for your report. You can adopt the poison pen method with structural formulae culled from the chemical literature or the Aldrich catalogue. You could also trace enlarged copies of these using a drawing pen, ruler and tracing paper. Reaction arrows can be drawn with a pen and lettering added to the master diagram in any of the ways described above.

Inorganic structures, which are less common in the literature, may need to be traced from your own drawn masters. If you are not typing your own report, you must discover well in advance if the system your typist will use can cope with your requirements. If it cannot, you will then have time to make alternative arrangements for the more specialized components of your report. What you must do is to show the location and size of any diagrams you wish to include on your manuscript so that the typist can leave the correct spaces for them.

7.7 PHOTOGRAPHS

For reasons of cost the publication of photographs in scientific journals is strongly discouraged. This attitude spills over into reports and you must normally have a very good reason to include any photographs in your project report. Photographs of apparatus, for example, are not sanctioned because it can be adequately specified in the text or shown in a line diagram.

Some of the only cases where photographs might be permissible include photomicrographs of catalytic surfaces or a series of photographs illustrating corrosion studies. Photographs of space-filling molecular models are at the bottom limit of acceptability. Nearly all these subjects can be photographed successfully in monochrome and it is unlikely that you will wish to include colour photographs in your report.

Unfortunately most commercial photographic establishments seem geared to producing colour photographs. It is becoming increasingly difficult to obtain good quality black and white prints of the type required for your report. Before deciding to include any photographs you should make sure that you can obtain good quality prints in which the range of tones varies smoothly from clear white to full black.

The finished prints should be either of A4 size with the same width white margins as the text (see Section 2 above), or be no wider than the text so that they can be mounted on a normal page of your report. This page must be typed or printed before the photograph is affixed since photographs cannot pass through laser printers or around rollers.

The larger A4 size prints can be bound into your report along with the other pages. Smaller prints can be fixed to an ordinary page with a photographic spray mountant or, if the facilities are available, fixed to the page with dry mounting tissue. Excess dry mounting tissue can be trimmed from the edge of the prints with a sharp blade. Excess spray mountant can be removed by a swab dampened with 40–60° petroleum ether.

Ideally the prints should be made on glossy surfaced, fibre based photographic paper. This gives better quality, more durable prints than can be obtained from resin coated paper.

In most cases the photographs will be taken and processed for you by the department's technical staff who will do their best to meet your requirements. Photomicrographs, particularly those taken with a scanning electron microscope, are usually of low contrast and require skilled printing. It is worth discussing methods of showing scale on photomicrographs with the technical staff. It is the author's belief that the scale should be shown on the original negative so that any subsequent enlargement or reduction is shown automatically and does not require calculating afresh each time.

If you are taking your own photographs they should be in sharp focus and clearly defined. The subject should fill the picture. You will probably be able to meet these conditions more easily by using a 35 mm SLR camera than with a compact camera. The former has a better viewfinder system and can accept medium focal length lenses (90–135 mm) which allow you to fill the frame with the subject from a greater distance. This greater camera to subject distance gives a more natural perspective. If you do use a compact camera, use the tele lens setting to achieve this effect.

For the best quality results use a slow black and white film (25 to 100 ISO) and avoid overdevelopment. It is a good idea to take several photographs of each subject, varying the exposure settings on your camera if this is possible.

Avoid cluttered backgrounds. Small objects can usefully be photographed on a sheet of graph paper to give a better idea of scale. Otherwise use a sheet of medium grey paper as a background. This colour avoids misleading through-the-lens exposure meters as would a black or white background. Further, these background colours blend with the black carbon or white hydrogen atoms of a space-filling molecular model.

Lay the subject near the front of the grey paper and curve the remaining paper up behind it. This avoids the sharp angle between the bench and the background from appearing as a confusing black line in the final print.

If you use flash as your source of illumination it should be diffused to avoid harsh shadows. Either bounce the flash onto the subject from a large white sheet of paper, which acts as a reflector, or place a piece of tissue over the flash window. In both cases make increases in the exposure to compensate for the weaker light that results. Use another large white sheet of paper on the far side of your subject to reflect light back into the shadows as a further means of reducing their otherwise inky blackness.

Photographic quality is usually destroyed by photocopying. If more than one copy of your report is to be produced, have the required number of photographic prints made and affix them to duplicate pages.

7.8 INSTRUMENT OUTPUT

The output from many instruments, particularly spectrometers, is often a graph. Unfortunately the chart paper used on many instruments is singularly unsuitable for inclusion in your report by direct photocopying. Attempts to reproduce it in your report may look nearly as bad as the infrared spectrum shown in Figure 7.7(a).

This figure deliberately shows many faults. First, the whole spectrum is reproduced when only part may be required to elucidate the point you are trying to make. Secondly, the standard panel for sample details has been retained instead of including any relevant details in the figure's caption. Most importantly, the reprographic process has shown the over numerous chart divisions almost as strongly in some places as it has the spectrum itself which was drawn in the traditional red ink. Elsewhere the chart divisions are weak or absent on the photocopy. There is no excuse for reproducing a spectrum overlaid with the diagonal stripe indicating that the end of the chart paper is imminent.

For reproduction in your report you should select the portion of the spectrum you require and draw it with black ink in the instrument's pen upon a plain, white roll of chart paper. Using your drawing pen you can then add suitable calibration marks upon the paper. You can add axis labels using any of the methods detailed in Section 7.5. The master diagram can then be reduced to the required size for inclusion in your report (Fig. 7.7(b)).

The digital output from modern spectrometers can be recorded and saved upon a computer disk for later manipulation or reproduction. Properly used, these instruments allow you the luxury of 20/20 hindsight. If you are using an earlier generation of instrument you must be prepared to change to a plain chart paper and a black inked pen whenever you think

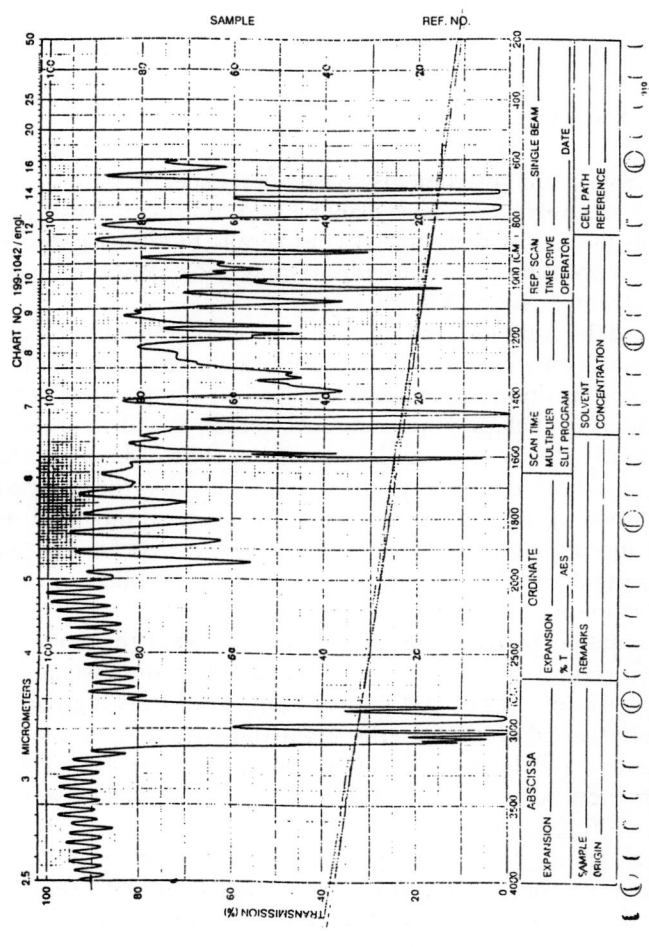

Figure 7.7 (a) Instrument output; (b) report version.

you have captured a spectrum worthy of reproduction in your report. It is better to record a few, eventually unwanted, fair copies of spectra than to have to repeat a reaction or whole preparation merely to record a spectrum of reproduction quality. If you should have to repeat your work for this purpose your only consolation must be that you are inadvertently checking the reproducibility of your work.

7.9 PROOF READING

It was noted above that the manuscript you supply to your typist should be as close to the final version of your report as possible. If you are typing the report yourself this is also a worthy objective since you will make fewer mistakes typing from an accurate manuscript.

In the main this accuracy can be achieved by carefully reading through the manuscript to eliminate the errors it contains. The process is known as proof reading and it is a skill you can soon learn. You will realize that you are becoming a skilled proof reader once you regularly find typographic errors in everything from newspapers to novels.

The reason you find these typographic errors is because you are reading the work of others. When reading your own work you have the fatal tendency to read what you wish to be there rather than what is actually there. It is very beneficial to have your text proof read by somebody outside the subject. The author's wife, a bacteriologist, finds innumerable errors in his chemical manuscripts simply by reading what is written and by not ignoring errors that a chemist, more immersed in the topic, would overlook. It may be possible for you to proof read the work of a student from another discipline in return for help with proof reading your own project report. Biologists call this process symbiosis!

Some word processing systems incorporate a program to check spellings. These are very useful but have their limitations. Some typographic errors give rise to valid words thus 'dread' becomes 'drear' by hitting 'r' on the line above for 'd'. Most importantly the programs' vocabularies, although large, do not include many chemical words. Although some programs make provision for additions to the master vocabulary, many chemical words are so similar that they can be mistyped and creep through if total reliance is placed upon the machine. Other programs allow through American spellings. Thus both words in the following pairs may be equally valid, litre (liter), colour (color) and behaviour (behavior).

It can be seen that there is, as yet, no substitute for human proof reading, although the programs do pick up many of the mistakes. However, proof reading is not solely concerned with the correction of spelling mistakes. While the programs can often tell if a word has been typed twice in succession, they cannot tell you which word has been omitted, nor can they point out that sentences begin with a capital letter. All these things are left to the vigilance of the human proof reader.

Even then there are other tasks for the proof reader who must ensure that cross references are correct. If the statement 'The reduction potential of aqueous bromine to the bromide ion ($E° = 1.065$ V) is greater than that for the reduction of ferric ion (see Section 4.7.3) occurs, then the proof reader must check that the value for the reduction of Fe^{3+} to Fe^{2+} does indeed appear in Section 4.7.3.

Other duties of the proof reader that merge with those of the sub-editor include checking that all chapters, references, equations, figures, tables and structures are numbered in their correct sequences. There are so many tasks to be accomplished that it is better to approach them methodically and use the following list.

Check-list for proof reading the manuscript

- Are all pages present, numbered and in the correct order?
- Are all spellings (particularly chemical words) correct?
- Are all formulae correct?
- Do all headings agree with the contents list in style, number and position?
- Do balanced equations actually balance?
- Are all alterations clear and unambiguous?
- Are all symbols clearly marked?
- Are all requests for different fonts clearly shown or implied?
- Is the punctuation correct?
- Are all tables, figures, schemes and structures mentioned in the text?
- Are all cross-references correct? (Always refer to sections, never to pages.)
- Have the spaces for artwork been indicated?
- Are all the sections and subsections numbered sequentially?
- Are all references numbered correctly?
- Are all names in the reference section spelt correctly?
- Are all equations numbered sequentially?
- Are all tables numbered sequentially?
- Are all columns and rows present in the table?
- Are all columns of the table correctly aligned?
- Are all figures numbered sequentially?
- Are all structures numbered sequentially?

Once the manuscript has been converted into typescript the whole process must begin again. It is to be hoped that you detected most of the errors in the manuscript because you will pay to have any residual error corrected in the typescript. However good your typist is some additional minor errors will have crept in while the manuscript was being typed. These will also have to be detected. Again, it is as well to be methodical about the whole process and go through the following check-list.

Check-list for proof reading the typescript

- Is the typescript legible? Is it black (not grey)? Has it become damaged in the printer? Are copies smudged?
- Are all words spelt correctly? Check particularly all chemical, foreign and proper names which may be outside the typist's or spelling check program's experience.
- Are words correctly capitalized?
- Are all chemical formulae correct? Check all subscripts, ionic charges and nesting of brackets.
- Has any passage or word been omitted?
- Has any passage or word been included twice?
- Are all special terms and symbols in the correct font?
- Are all headings in a correct and consistent style?
- Is the report in the correct order?
- Are the pages numbered correctly?
- Has space been left for any characters or symbols you must insert yourself?
- Have the correct spaces been left in appropriate places for artwork to be included?
- Are there any bad line or page breaks that require correction?
- Are the margins correct?
- Are all lines correctly spaced?
- Are the paragraphs inset?
- Is the punctuation correct?
- Are all the sections and subsections numbered sequentially?
- Are all references numbered correctly?
- Are all equations numbered sequentially?
- Are all tables numbered sequentially?
- Are all figures numbered sequentially?
- Are all structures numbered sequentially?
- Are tables correctly presented? Do columns of figures align on the decimal point? Has any column or row been omitted?
- Is all the text in the correct font? (Pay special attention where the font has been changed legitimately.)

Graphic supplies

Most university towns have graphic supplier's shops where you can buy drawing pens, ink and tracing paper. In case of difficulty, drawing requisites can be obtained from Oram and Robinson, Cadmore Lane, Cheshunt, Waltham Cross, Hertfordshire, EN8 9SG.

Specialized chemistry stencils can be obtained from the Aldrich Chemical Co. Ltd, The Old Brickyard, Gillingham, Dorset, SP8 4BR.

The project talk | 8

In later life you will almost certainly be required to talk about your work before a formal gathering of your peers. The lecture on your project work may well be your initial experience of doing this. In many ways this is unfortunate, since you will be performing before a knowledgeable audience who earn their daily bread by giving lectures themselves. When examination stress is added to these factors, it is not surprising that many students regard a lecture on their project as a baptism of fire.

This chapter will show you how to increase your confidence by improving your performance. The improvement can be made by considering the lecture's content, the style of delivery and the use of visual aids.

8.1 BUILDING YOUR CONFIDENCE

The main thing you have to fear when talking about your project is fear itself. If you are over-anxious about lecturing to the assembled staff and your fellow students then your performance will suffer, and an autocatalytic cycle will be initiated to the detriment of your presentation. It is, therefore, very important to prevent your anxieties from feeding upon themselves. Anxieties are cowards and can often be vanquished merely by confronting them. Some of the more common sources of anxiety are listed below.

- Looking a fool
- Making a mistake
- Omitting something of importance
- Being unable to answer questions.

It is of little comfort to realize that many of these anxieties are both irrational and unlikely to occur. However, if you spend your lecture waiting in dread for a calamity to happen, then it is likely that it or another less common blunder will occur.

If the sources of each anxiety are investigated it is possible to take countermeasures against the anxiety and banish it. By far the most common anxiety is 'looking a fool' which is a blanket term for several associated anxieties. These stem from a lack of self-confidence and include 'I look a mess', 'I have a funny accent', 'I gabble when under stress', 'I will talk too loud/too softly' and 'I will suddenly be lost for words'. Some of these fears are not totally irrational except that the most beautiful or handsome students frequently believe that the audience will home in on the one minor flaw in their appearance.

You will gain confidence by appearing at your best when presenting your lecture. This is not to say that you should be done up like a dog's dinner to the detriment of your comfort. However, there is no need to gratuitously insult your audience by appearing in ragged or unwashed clothing. Similarly, males sporting designer stubble or females with the latest spiky, green hairdo will alienate some of the more fuddy-duddy senior academic staff.

A neat, clean and not overly flamboyant appearance should be the aim. If you feel comfortable and know that you have avoided the wilder excesses of dress and appearance, then there is little reason to believe that others will think you look foolish.

Another way of looking a fool is to have to dash to the toilet in the middle of your lecture. Always visit the toilet immediately before you give your lecture, because anxiety can speed up some physiological processes. Never encourage visits to the toilet by partaking of alcoholic drinks to 'steady your nerves' before your lecture. If you read the final chapters of Kingley Amis's *Lucky Jim* you will see how alcohol, far from improving your performance, can seriously detract from it.

Avoid irritating your audience by some persistent mannerism or phrase. Good examples of the latter can be heard when sportsmen are interviewed. Never put your hands to your face, or commit such breaches of accepted manners as indulging in a scratching session when you are speaking. Do not pace about while speaking. This distracts the audience from the content of your lecture.

A suitable lectern is of valuable assistance when delivering a lecture. Placing your notes upon it stops you from looking down onto the table when reading them. Holding onto its sides anchors you in place and stops you worrying about what to do with your hands. Additionally this posture helps you project your voice to the back of the room. However, avoid clutching it rigidly throughout your lecture or you will look as if you have been welded in place. It may be necessary to move slightly to present your visual aids. Site your projector near the lectern otherwise your lecture will degenerate into a series of route marches between the two accessories.

Many students mistakenly believe that a pronounced regional or national accent will be a handicap when they are talking about their project. This is quite untrue but an attempt to overlay it with standard

English will leave you paying less attention to what you say than to how you are saying it. This is a sure recipe for disaster. While many find accents attractive, regional colloquialisms are not usually suited to conveying advanced scientific thoughts and should be avoided.

The remaining anxieties can be countered by careful preparation. It is always good practice to prepare the text of your lecture in advance and read it to a tape recorder. As you listen to the playback you should be able to identify those parts of the text that could be improved. Try to speak at an even pace throughout like the newsreaders of radio and television. It will seem measured, or even slow, but that is better than an incomprehensible gabble. The duration of the taped performance will tell you if the script of your lecture needs pruning or if you have time to go into more detail.

Do not forget to allow time for the audience to assimilate your slides or demonstrations. Every slide needs a minute's talk, so it is better to present a few good quality slides than many bad ones. Demonstrations will absorb much more time and are even more accident prone. They are correspondingly impressive if carried out faultlessly and with aplomb. Practice is **essential**.

A few performances of your lecture, first to yourself via the tape recorder and later to your friends, will increase your experience and boost your confidence. Small groups of students make good trial audiences for each others' lectures.

A script for your lecture is a great help when you are first speaking to an audience. The most important thing to remember is that your lecture is not a public reading of your project report. Your script should be in a different style to your report and allow more interaction with your audience.

If you are using slides these also serve as an outline script. You can use the slides as reminders and time markers while you are speaking. Although scripts destroy spontaneity, they do act as a safeguard against omissions and mistakes.

It is very simple to avoid talking too loudly or too softly during your lecture. An earlier visit with your fellow students to the lecture theatre or room where the talk is to be given will allow you all to practise your delivery. You will find that even large lecture theatres do not require you to shout at the volume normally reserved for abusing referees. Provided you speak with your head thrown slightly back, your projection will allow you to be heard at the back. Your audibility will be improved by stressing the last syllable of each word slightly more that you would in ordinary conversation. Never hang your head down and mumble into your lecture notes. This will ensure that you are inaudible to anyone beyond the third row. Always face your audience when speaking. Never mutter incantations to the blackboard when talking about the writing on it.

Arrange a system of hand signals for too loud/too quiet with a friend.

Ask your friend to sit at the back when you give your lecture and act as your volume control. If your friend can also smile at you occasionally during your lecture this will improve your self-confidence.

Being unable to answer questions is a frequent anxiety. It does have some foundation, but usually the only person in your audience who knows more about your project and its subject will be your supervisor. If you have remained on good terms with your supervisor you should have nothing to fear from this quarter. The most likely source of questions will be any errors in your presentation. Members of staff who are used to proof reading and marking examination scripts are very adept at locating these errors, particularly when they occur on slides or transparencies. In Section 8.3 you will be advised how to detect errors in your visual aids before projection.

Students would be less worried about the questions after their lecture if they knew how difficult it can be to devise intelligent and relevant questions following even a competent lecture on a well constructed project. This is why the staff have to resort to drawing your attention to the minor blemishes mentioned above. If, from the corner of your eye, you see the chairman scribbling away furiously during your delivery do not be alarmed. It is more likely that a possible question is being noted than that you have made a serious blunder.

That is not to say that perceptive questions cannot arise from most unlikely sources. The author recalls an inorganic chemist, whose wife was a haematologist, severely testing a student's knowledge of the relationship between blood-iron levels and obscure medical conditions.

To create a better impression, and to guard against such eventualities, you should read around your topic thoroughly. Obviously you cannot be expected to know everything and it is permissible to reply 'I am sorry, I don't know about that aspect' on one or two occasions during question time. Repeating this formula more frequently will certainly damage your marks, but probably not so severely as digging a huge pit for yourself under astute cross-questioning.

The wider knowledge you have gained from reading around the subject can often be used to parry questions by showing you are aware of a related aspect of the subject. Although it is unethical, you could always plant one or two questions amongst your colleagues in the audience. These favours can be returned when it is their turn to speak on their projects.

The most difficult questions to answer are not the intelligent ones but the stupid ones. However lucid your presentation has been there will always be one member of your audience who has totally misunderstood it. Unfortunately these people often feel compelled to demonstrate this fact to the others present by asking a stupid question. Faced with such a question you not only have to look at your whole project from a totally different angle, but simultaneously devise a civil and rational answer. Perhaps the best

ctic is to ask for clarification of one part of the question. Then you will
ain time to think and the questioner instead may become confused.

Answering questions is a valuable part of your training. It should not be
ooked on as an ordeal but as an opportunity to demonstrate your
nowledge of your project.

2 WRITING THE SCRIPT

is vital to remember that your project lecture is not a public reading of
our project report. One of the most important differences between the
vo is the combination of the experimental and results sections in the
cture. The audience will wish to know how you obtained each result
efore you discuss its significance. However, both your project report and
cture will benefit from having an authoritative and concise introduction.
hese may not differ too greatly in style, although only the most important
atures of your introduction can usually be included in your lecture
ecause of the limited time you have available.

Nevertheless it is important that you do not present facts too rapidly for
em to be assimilated by your audience. You can dilute the flow of facts
' including more detail than is customary in a written report. One
ample that springs to mind is a more detailed description of what
ppened during a reaction. In your report you normally only give the
itial and final colours of your solutions, in your lecture you can describe
e rate and sequence of colour changes observed. You can also tell your
dience more about the difficulties you encountered and how you
ercame them.

This latter feature illustrates another essential difference between a
port and a lecture. Reports are read to retrieve the facts, but lectures can
ly convey these facts successfully if you gain a rapport with your
dience. Involving the audience in your own difficulties and excitements
kes them more receptive to the ideas you are trying to convey. Some
turers are able to use humour to win their audience over to their side.
careful of this tactic: it is easy to misjudge the mood of your audience. If
u consider yourself an amateur comedian then strictly ration your jokes.
pun usually avoids giving offence and both bad and good puns seem to
equally memorable.

Although you should have developed your arguments logically in your
port, it is even more important that you do so in your lecture. Your
dience cannot refer back to an earlier page to help them follow your
guments. Concentrate on one theme at a time and keep the argument
id and simple.

Try and avoid hypotheses based purely upon numbers even if you are
aling with numerical material. Numbers are very difficult for the

audience to remember while they are listening to you talk. Rely on trend if you can. State that a bond is shorter in one compound than another rather than giving the exact bond length in each case. This would require your audience to remember both values and carry out mental arithmetic during your lecture! If the exact numerical values of the property are very important state them on a slide or transparency and keep this in view while you discuss the significance of the values. In this way members of the audience do not have to rely upon their memories. Similarly, structures are difficult to carry in the memory when listening to a continuing theme. structural features, which are often difficult to describe, are important show the structure on a slide while talking about it.

Above all, try and sound committed and enthusiastic about your project Audiences detect a lack of commitment before they detect a lack of ability If you sound half hearted about your project then the staff may be subconsciously half hearted about awarding you marks for your lecture Always accentuate the positive aspects of your work. Say how pure your intermediate was to disguise the miserable yield (or *vice versa*!). Play down the mundane aspects of your results and concentrate upon the novel features you have unearthed.

Continue this positive attitude into your conclusion. Do not be afraid state boldly what you consider you have achieved. Ending on a defeatist note leaves the way open for a sadistic questioner to turn it into rout. By ending on a positive note you make it much more difficult for a damning question to be asked.

8.3 VISUAL AIDS

Visual aids enliven any lecture provided that they are relevant and legible to the back row of the audience. The simplest visual aid is the chalk mark on the blackboard. When writing on the blackboard make sure that you write boldly enough for those in the back row to read what you have written. Do not write large sections of your lecture on the blackboard. You are not talking to a primary school class too young to take their own notes. Long periods spent in silence with your back to your audience is not a good style to adopt. Involved diagrams must be prepared in advance. This requires more advanced visual aids whose images can be projected upon screen.

These images may either be on 35 mm film (slides), or on overhead projector transparencies. The latter are easier to produce, more versatile and can be seen by your audience in an undarkened room. Flip-charts may be used for small audiences, but assessments are usually carried out before larger groups where this type of presentation is unsuitable. However, if type of visual aid you select will probably be dictated by the facilities available in your department. You should ascertain well in advance of the lecture what visual aids can be used.

You should also find out if the method you intend to use will be appropriate to the room that it will be used in. Slides normally require a blacked out room. The smaller magnification of overhead transparencies allows them to be shown in ordinary illumination. However, a bright, sunlit room will result in degraded images and will only be suitable for the boldest and most simple transparencies. If the room is equipped with curtains make one of your friends responsible for drawing them when you show transparencies.

Whatever method of projection you use, your original artwork should be of the highest quality since any imperfections will be greatly enlarged upon the screen. Laser printers are ideal for preparing text or the labels for graphs' axes. However, many computer graphics packages and structural formulae programs are still too crude to give acceptable images. You may still need to draw out these parts by hand using a drawing pen.

Transparencies for overhead projectors are easily made by photocopying A4 originals onto transparent film. Photographic slides cannot be prepared so simply since most cameras cannot focus closely enough to fill the frame with an A4 original. You will probably find it best to photograph A3 originals to obtain 35 mm slides. Since most typewriters and graphic aids are geared to the A4 size this adds to the difficulty of producing the original artwork. Very few students have the photographic skill or equipment to produce acceptable 35 mm slides. However, anybody who can operate a copying machine can produce an overhead transparency from original artwork.

The full and accurate range of colours obtainable is the principal advantage of 35 mm slides. Most overhead transparencies are reproduced as black on clear, although coloured bases are available. Colour can be added by hand to overhead transparencies using coloured pens. Few departments have the facility to produce coloured overhead transparencies although these can be made commercially at fabulous expense in most cities.

The most common mistake you can make is to try to include too much information on each slide. A trip to the back row of the lecture theatre will soon show you that complicated reaction schemes or large tables are unsuitable subjects for slides. You can avoid this common mistake by taking a sheet of A4 paper and, using a thick felt tip pen, writing out your text. The maximum quantity of information per transparency is all you can write on the A4 sheet in this way.

Avoid detail and concentrate the main points. Do not add verbiage to the transparency's text. Say 'First-order kinetics' instead of 'Let us next consider first order kinetics'. Never use transparencies showing normal size typescript: this will look like the type size newspapers use to report the croquet results when it appears on the screen. Always enlarge typescript to at least 7 mm high on the transparency to ensure legibility. Ten or twenty millimetre high lettering is even better.

Tables can usefully be replaced by graphs, since not only are too many data in a table confusing, but your audience will not have time to assimilate them. Tables, with their greater precision, may play an important part in your written account where your readers can examine and compare the information presented at their own speed. However, in your lecture you will be dictating the rate of information presentation.

Reaction schemes are best broken down into a few steps per slide. After all, you can only talk about one step at a time.

The greater versatility of the overhead transparency arises from your ability to add to or subtract from the information it presents at will. Information can be subtracted simply by blanking off the appropriate portion of the transparency with an opaque sheet. This is very difficult to do with a 35 mm slide.

Do not allow your audience to read the entire transparency while you search for a piece of opaque paper, and avoid knocking the transparency askew when blanking off part of it. Instead, improve your presentation by pre-positioning the opaque paper. This paper can be fixed to one side of the transparency mount with a piece of adhesive or masking tape as shown in Figure 8.1(a). If you have a choice, the latter is better because it does not weep adhesive on heating and literally gum up the works. Alternatively use an overlay by making the second point on a second piece of transparent base which can be fixed to the side of the master transparency's mount in the same way (Fig. 8.1(b)).

Transparent overlays are very useful for making sequential points (e.g. steps in a reaction mechanism). You should not attempt to use more than three overlays on a transparency lest you commit the cardinal sin of placing too much information on one transparency. You will find it is easier to manipulate the overlays if you fix tabs (to help lift them) on the opposite side from the hinge. In the hot, dry atmosphere above the projector platen, static electricity causes the original transparency to adhere to overlays. Take care that your overlays, which hang down from the projector platen when not in use, do not obstruct the flow of cooling air to the projector's powerful bulb.

The overlays should be drawn on a sheet of tracing paper that has been fixed over a copy of the main transparency. Used intelligently this method should ensure the correct register of any overlay. Where very accurate register of an overlay is required draw a cross with a fine pen in the same corner of each tracing and make sure these are in exact register before starting on each overlay.

Always provide yourself with several blank acetate sheets and marker

Figure 8.1 Increasing the versatility of transparencies for the overhead projector.

pens for them before giving your lecture. You can then make additional points, possibly in response to questions, upon these spare sheets without defacing your original transparencies which they overlay.

You can increase the impact of monochrome transparencies by highlighting the main points with colour. You can add this with a suitable felt tip pen. However, you may only wish to highlight your current point. You can do this by placing a piece of transparent, coloured Perspex (polymethylmethacrylate) over the item. Small pieces of fluorescent red or green acrylic sheet about 3 mm thick can be obtained as cheap offcuts from the manufacturers of shop signs and fascias. You can cut these pieces to the desired shapes with a hacksaw. The rough sawn edges are smoothed successively with a file and decreasingly coarse grades of glasspaper. Take care that you do not scratch the soft facing surfaces of the acrylic sheet when working with the material.

A group of you could easily make up a communal set of acrylic shapes for your lectures. Your kit of prettily coloured pieces of acrylic sheet should include a variety of squares and rectangles. Long rectangles of contrasting colours can be laid over the axes' labels on graphs to emphasize them (Fig. 8.1(c)). A 15 mm square can be used to overlay a single character on the transparency. Thus it can be used to highlight the number of an item in a list. Rectangles can be used to highlight words, and very thin rectangles can be used to underline titles to increase the impact of monochrome transparencies (Fig. 8.1(d)). If you attempt to use very small pieces of acrylic sheet you will encounter problems from static electricity which will prevent you placing them accurately.

It is very useful to make a pointer from the acrylic sheet. You should make this about 150 mm long and no more than 10 mm wide. Evenly taper one end of the original rectangle to a point. This pointer can be laid on a transparency to indicate the item under discussion, for example a peak in an infrared spectrum (Fig. 8.1(e)). Lay the pointer upon the transparency, do not move it and irritate your audience by waving it about while you are speaking. Thin knitting needles are better pointers than pencils, which in turn are infinitely superior to fat, stubby fingers.

Never forget that you can project the image of any coloured, transparent object on the platen of an overhead projector. The internal reflections of small crystals give rise to black images on the screen, but if you dissolve the crystals in a non-flammable solvent contained in a small Petri dish or beaker then this can be placed on the platen and projected. It is even possible, if you exercise great care, to show colour changes during a reaction by carrying out the reaction in a beaker on the platen. Never spill liquids on the projector unless you wish to bring your lecture to an expensive and spectacular end. Never use corrosive liquids on the projector. In case of disaster switch off the projector immediately and have a kitchen towel to hand to mop up any spillage.

You must carefully proof read the art work before making the final transparency or slide. This is difficult since diagrams do not lend themselves to systematic scrutiny. For example a branch in a mechanistic scheme can lead to part of it being overlooked. If you have stencilled the lettering for the diagram or graph it is very easy to suffer from sign writers' disease and omit letters or add them twice. The worst example the author ever saw was one he prepared for an American undergraduate class. This had 'litre' and 'liter' randomly alternated throughout the slide. Fortunately this was so bad that the mistakes were detected before producing the transparency. It is much more difficult to detect a single mistake. The problems of finding a single mistake in a complicated diagram are a further incentive not to include too much detail.

Check-list for proof reading artwork
- Are all words spelt correctly?
- Are all formulae correct?
- Do equations balance where required to do so?
- Have pieces of display formulae been omitted?
- Is the slide well arranged and balanced?
- Has any extraneous information been included? (Prune your material ruthlessly.)
- Are there any smudges or other blemishes present?
- Are all graph axes labelled and scaled correctly?
- Can overlays be placed in correct register?
- Does each scheme and graph bear a title?

8.4 DELIVERING YOUR LECTURE

If you have followed the advice given in the previous sections you should have written your script, prepared your visual aids and checked the timing of your lecture. On the day of your lecture you should be comfortably dressed and have with you your script, visual aids and other sundry accessories. If you are using slides these should be correctly spotted and placed in the correct order before handing them to the projectionist. If you are using overhead transparencies these should also be placed in their correct order. Do not forget to take spare acetate sheets and pens to write on them, acrylic highlights and any other visual aids you wish to use.

Immediately before your lecture you should check that there is a pointer, a blackboard eraser, sufficient chalk in a variety of colours and that the blackboard is clean. Look at all surfaces of roller boards lest some practical joker has written a derogatory message on a surface you will bring into view during your lecture.

Pay similar attention to projection apparatus. Check that it is in working order and that the optics are clean. An efficient department will have a spare projector standing by.

If you are to operate the projector yourself you should know how to do so. Learn before your lecture, not during it. If you have made preliminary checks on the effectiveness of your visual aids in the same lecture room, you will have learnt how to operate the apparatus, and where the plug points and light switches are. Make sure that the projector is at the correct distance from the screen to give an image of the desired size. If the image is too small, move the projector away from the screen. If it is too large move the projector towards the screen. If you are using both portrait and landscape format slides ensure that both will fit on the screen.

It is also important to ensure that the projector is square on to the screen. If one side of the image is smaller than the other, move the projector towards the other, larger side. Unfortunately many projectors are pointed upwards to throw an image on a screen higher than the projector. This results in a 'keystone'-shaped image. It will not usually be possible for you to raise the projector sufficiently to obtain an orthogonal image.

In severe cases the keystone effect will make it impossible for you to focus sharply on the entire slide simultaneously. The best compromise is to focus the projector sharply for one third of the distance from the top of the image. You may need to alter the focus as you progress down your slide. This is irritating for both you and your audience but it is better than having the audience strain their eyes.

If you are using slides instead of overhead projector transparencies, you may require the services of a projectionist. To help your projectionist, have all your slides in the same type of mount. Glass mounts are better, since the slide will not 'pop' out of focus as the film base changes curvature as it dries out in the heat of the projector's lamp.

Arrange your slides in the correct order before giving them to your projectionist. Have the first slide on top of the stack. Draw a diagonal line with a broad, felt tip pen across the side of the stack once it is in order. This makes it easy to see if a slide has become displaced or is missing.

The slides must also be correctly oriented before projection. You should place a coloured, 5 mm spot on the bottom left hand corner of the mount when the slide appears as you wish it to be seen on the screen. Slides should be placed in the projector with this spot in the top right hand corner facing away from the screen. These spots can be purchased from stationers or photographic suppliers. However, it is cheaper to make them from coloured sticky labels using a standard 80 mm punch.

You must agree on a signal with your projectionist when you are ready for the next slide. It may be simply announcing 'Next slide, please' or tapping your pointer on the floor. To avoid confusion make your slide series sequential. Do not ask the projectionist to return to a slide during the course of your lecture. If you want to show a slide a second time have a copy made and insert this into the sequence at the correct point.

If you are projecting the slides yourself, spare sufficient time to master the remote control device. Make sure that you can refocus smoothly, and know which button to press to move forward through your slide sequence. Remote control projectors benefit more than most from a homogeneous set of thin, glass-mounted slides.

Check-list for your lecture

- Does your material fit the time allowed?
- Have you rehearsed your presentation?
- Can your visual aids be projected/displayed?
- Can your visual aids be seen clearly in the back row of the lecture theatre?
- Do you know how to operate the projector(s)?
- Are the projectors aligned and focused on the screen?
- Are the illumination levels of the room and screen images compatible?
- Is the lectern in the correct position?
- Do you have the following with you?
  Your script
  Visual aids
  Acrylic highlighters
  Pointer
  Coloured chalks
  Board eraser
  Spare acetate sheets and ohp pens
  Watch or clock
  Your project report for reference
- Is your script in the correct order?
- Have you visited the toilet?

9 | Poster presentations

Posters play an increasingly important part at chemical conferences, not least because acceptance of a poster often means a reduction in conference fees for impecunious postgraduates. Nevertheless, posters are a useful way of imparting recent results and making valuable contacts at conferences. Accordingly many chemistry departments ask students to present their project results in poster form. The more enlightened departments even go so far as to imitate the custom of international conferences and offer a bottle of aqueous ethanol as a prize for the best poster.

It is to be hoped that your department will not follow that other custom of international conferences and fail to disclose the size of the display boards for each poster. You should find out the size of your display board as early as you can so that you can design your artwork to fill this area as neatly as possible.

When you prepare your poster you should remember the golden rule that you applied to your slides or overhead transparencies – **don't include too much detail!** All that is required is a clear, brief exposition of what you have achieved. People can read your poster at their own pace in contrast to a project lecture which is given at your pace. Therefore, it is permissible and even desirable to include spectra, tables, reaction schemes or crystal structures, since these enable you to break up the passages of text into small, comprehensible blocks.

Once upon a time producing these blocks of text was a daunting task for all but the most artistically gifted students. Fortunately the development of the laser printer and the enlarging copy machine has made it possible for anyone to produce a visually cohesive poster. Have your text printed out by the best quality laser printer you can find, since any imperfections will be magnified upon enlargement. You can also use the laser printer to prepare captions for figures and to label the axes of graphs.

Depending upon the quality of your word processing system you may also be able to produce diagrams and chemical structures via the laser printer. Again apply the rule 'If you can tell it was done on the computer it isn't good enough', and be prepared to draw these items out by hand. Only the most advanced systems will provide you with coloured diagrams, so you should give some thought to increasing the appeal of your diagrams by making provision at the design stage for later manual application of colour.

Including colour may mean that the diagrams you used in your written report will need modifying before using them in your poster. If you are colouring a diagram with crayons or pens, use regular hatching running in one direction only. If you are using a paint brush, make sure that the ink or paint is sufficiently dilute. An even application of colour is best achieved when **all** the area being coloured remains damp while being painted or colour-washed. Mix sufficient paint or ink. Do not make the elementary mistake of running out of material half-way through colouring.

It is very easy to decrease the visual appeal of your poster by using too many presentational gimmicks. You should aim for a dignified style. A single typeface throughout is much better than using every font on your printer simply because, like Everest, they were there. Audiences are surprisingly critical of style. The author recalls a poster of his which was dubbed the 'UNESCO poster' merely because of an overlavish application of coloured column charts. He was saved from total ignominy by a neighbouring French poster which used aluminium foil hexagons to denote benzene rings. This was named the 'Playschool poster'.

Each poster requires a bold heading stating the title of your project and your name. Find out if it is departmental practice to include your supervisor's name also. If this is the case, place the supervisor's name last and follow it with the asterisk which denotes the senior author. Some departments produce these title boards for each project, in others it is your responsibility to set the scene for your project. If the remainder of your poster is on white paper, have the title printed on coloured paper to emphasize it.

Once you have all your artwork prepared, try to find out how it may be fixed to the display board. If you have mounted the diagrams upon boards make sure that you have enough long pins to fix them in place.

The presentation of your poster to the public gives them an opportunity to ask questions about the work you have described. This should present you with few problems, particularly if you have already delivered your project lecture. Nevertheless, you should have a copy of your project report to hand to answer any very specific queries.

When standing by your poster you should remember that you too are on display. As mentioned in Chapter 8 it is desirable to avoid the wilder flights of fashion on these occasions. Who knows if one of the visitors may wish to employ you on the strength of your project work?

10 | Applying your project skills

There is a sense of anticlimax after you have completed the last part of your project assessment, and it is all too easy to put your slim project report away in a drawer. You may resolve that nobody, not even a doting mother, will read it again. This is a very negative attitude to adopt towards something that has taken up a large slice of your life during the past year. Instead your project report should be regarded as an asset.

Will you be attending any job interviews in the near future, or are you seeking a research post at another university? You should take your report to these interviews since there may be an opportunity to produce it as evidence of your abilities. These opportunities are particularly likely to occur if your interview is informal and your interviewer is able to talk to you alone for any length of time.

Industrialists will welcome evidence that you are able to write a report. They will be grateful that they can teach you more productive things than report writing, even though they set great store by progress reports. Indeed, the experience you have gained may make your reports stand out and bring you to the attention of your new overlords much more rapidly than might otherwise have been the case.

Similarly another university department will look more favourably upon a candidate who can provide evidence of successfully completing an undergraduate project. It is even possible that you have chosen to apply for a research post in the same field as the project which fired your enthusiasm for research. More broadly, your project may have made you decide on a career in research rather than the service industries.

Many Masters degrees now also involve some project work. You will soon realize that the tricks you learnt completing your first project can easily be applied to the next, and that the second project will be much easier than the first. With a little refinement many of the lessons you learnt will also be useful when writing your PhD thesis.

Your report and the few overhead projector transparencies left from your lecture may be your only tangible reminders of your project.

However, you will continue to draw benefit from the intangible aspects of your project for many years to come.

You should have learnt to use the library more effectively after preparing your project report. Your project may also have made you aware of the potential of the primary literature. This will allow you to improve your lecture notes by investigating important topics in more detail and incorporating some recent advances and applications. You will find it is a surprisingly effective method of revision to write out more comprehensive and up-to-date lecture notes. By using suitable original examples to illustrate your examination answers you must improve your standing in the eyes of the examiner. Your better library skills will improve your performance in tutorials and the benefit you gain from them.

It is to be hoped that one of the most important things you learnt while doing your project was how to use your time effectively. Making good use of every little moment will be of paramount importance in the rup up to finals. Later you may have the opportunity to organize your own section in accordance with the organizational skills you first learnt when organizing your project schedule.

When writing your report you learnt how to rank and organize topics. Surely you can do this again when answering questions in your final examinations? If nothing else, the pressure of completing your project on time will have prepared you for the pressure of the final examinations.

While a successful project must increase your self confidence, even a less successful project can provide clues to areas of weakness. When revising for finals, remember that you are not really working if you only revise the things you know. Real progress is made by improving your performance in substandard areas. One word of warning – if you should find an examination question which asks about your project topic do not spend the entire examination writing about it. A perfect answer will only gain the maximum marks for one question. Equally, do not attempt to pervert other questions into questions about your project. This is only marginally less reprehensible than trying to pervert questions to correspond with last night's revision.

It may be some time before you can develop the minor managerial skills you learnt when you employed a typist and worked together as a team. However, you will frequently have occasion to remember and apply what you learnt in that exercise. This may be later in your career or even in your private life when you engage an architect to propose alterations to your mansion.

It will also be a long time before you are likely to address an audience as distinguished as that which attended your project lecture. Public speaking before lesser audiences should now hold no terrors for you. Once you have reached the stage when you cannot truthfully use the opening words 'Unaccustomed as I am . . .' you will become quite blasé about the prospect of speaking in public.

You should be able to look back on your project with pride and find many opportunities to apply the knowledge you have gained. Perhaps one day, far in the future, when you are congratulated on a good performance in some sphere you will be able to reply,

'It was nothing really – I learnt to do that when I was working on my student project in chemistry.'

Index

Appendix 1

GREEK ALPHABET

Upper case	Lower case	Name
A	α	alpha
B	β	beta
Γ	γ	gamma
Δ	δ	delta
E	ε	epsilon
Z	ς	zeta
H	η	eta
Θ	θ	theta
I	ι	iota
K	κ	kappa
Λ	λ	lambda
M	μ	mu
N	ν	nu
Ξ	ξ	xi
O	o	omicron
Π	π	pi
P	ϱ	rho
Σ	σ	sigma
T	τ	tau
Y	υ	upsilon
Φ	φ	phi
X	χ	chi
Ψ	ψ	psi
Ω	ω	omega